1976

Kinetics of Inorganic Reactions

Kinetics of Inorganic Reactions

by

A. G. SYKES, B.Sc., Ph.D.

Lecturer in Inorganic Chemistry
The University of Leeds

PERGAMON PRESS

OXFORD · LONDON · EDINBURGH · NEW YORK
TORONTO · SYDNEY · PARIS · BRAUNSCHWEIG

Pergamon Press Ltd., Headington Hill Hall, Oxford
4 & 5 Fitzroy Square, London W.1
Pergamon Press (Scotland) Ltd., 2 & 3 Teviot Place, Edinburgh 1
Pergamon Press Inc., Maxwell House, Fairview Park, Elmsford, New York 10523
Pergamon of Canada Ltd., 207 Queen's Quay West, Toronto 1
Pergamon Press (Aust.) Pty. Ltd., 19a Boundary Street, Rushcutters Bay, N.S.W. 2011, Australia
Pergamon Press S.A.R.L., 24 rue des Écoles, Paris 5[e]
Vieweg & Sohn GmbH, Burgplatz 1, Braunschweig

First edition 1966

Reprinted 1970

Library of Congress Catalog Card No. 65-27386

Printed in Great Britain by Bell & Bain Ltd., Glasgow, and reprinted lithographically by A. Wheaton & Co., Exeter

08 011440 7 (flexicover)
08 011441 5 (hard cover)

CONTENTS

CONVERSION TO S.I. UNITS

Activation energies and entropies of activation are expressed throughout in kcal/mole and cal/mole/°C respectively. These may be converted to S.I. units (kJ/mol and J/mol/K) by multiplying by 4·184.

Other conversions which are relevant to this text are 1mμ $\equiv$ 1 nm and 1Å $\equiv$ 100 pm

PREFACE

In this book I have tried to give a reasonably comprehensive and up-to-date account of the kinetics and mechanisms of inorganic reactions. Two main fields are covered, those of homogeneous gas-phase reactions, and solution reactions, the latter being almost exclusively in aqueous solutions.

Although many of the gas-phase reactions were first studied some years ago, recent more rigorous investigations have in a number of cases proved informative and well worthwhile. In particular one might mention here the decomposition of nitrogen dioxide and the reaction of hydrogen with iodine, reactions which are now known to proceed at least in part by radical and atom paths respectively. In recent years, with the development of new techniques, it has also been possible to study atom-recombination reactions in the gas phase.

Interest in solution reactions has been particularly marked during recent years. Such reactions fall into two main categories, oxidation–reduction processes in which there is either electron-transfer or atom-transfer, and substitution reactions, generally ligand-substitution reactions. There are also a number of acid–base reactions in which there is transfer of a proton but no change in electron structure or molecular configuration connected with the transfer. In addition to chapters on electron-transfer reactions between metal ions there are chapters dealing with the reactions of metal ions with diatomic molecules, solvent water, and perchlorate ions. The reactions of compounds of non-metallic elements such as hydroxylamine, hydrazine, hydrogen peroxide and sulphite ions are also considered, and there is brief reference to the recently studied reactions of the hydrated electron.

The following abbreviations have been adopted in referring to reactions involving " uncomplexed " metal ions in solution. With for example the ferric ion the formula Fe^{III} has been used to refer to the total ferric present, i.e. $Fe^{3+} + FeOH^{2+}$ . . . etc., it being understood that Fe^{3+} is in fact the hexaquo ion $Fe(H_2O)_6^{3+}$, and $FeOH^{2+}$ is $Fe(H_2O)_5OH^{2+}$ etc.

The book is written in the first instance for the senior undergraduate who has already followed a first course in kinetics, such as might be found in the larger physical chemistry textbooks, and who seeks a general account of inorganic reactions. At the graduate student level it is hoped that the references given will provide an adequate contact with the literature.

Finally I would like to thank Professor Irving for allowing me to research and lecture on kinetics in Leeds, Professor Irving and Dr. Higginson for their constant help and encouragement in preparing the manuscript, and my wife for typing a difficult manuscript.

Leeds
September, 1965

CHAPTER I

INTRODUCTION

CHEMISTRY is concerned not only with the properties of elements in their combined and uncombined states, that is with substances at equilibrium, but with the nature of transitions between equilibrium states. Kinetics is the study of the speed and manner of all such transitions.

The mechanism of a reaction

The first aim in kinetics is to identify the basic elementary reactions, i.e. those reactions which cannot be resolved into a series of simpler steps. Very often, the overall stoicheiometric equation gives no information as to the reaction sequence or mechanism. Thus the fast reaction

$$MnO_4^- + 8H^+ + 5Fe^{2+} \rightarrow Mn^{2+} + 5Fe^{3+} + 4H_2O$$

cannot proceed by the simultaneous collision of fourteen reactant ions (which is highly improbable), but must proceed in a stepwise manner with the intermediate formation of different oxidation states of manganese. Just which oxidation states are involved in such a complex system as this is difficult to establish however. Much simpler overall equations,

$$H_2 + Br_2 \rightarrow 2HBr,$$

are not always, themselves, representative of the elementary reactions and in this particular case the mechanism has been shown to be

$$Br_2 \rightleftharpoons 2Br$$

$$H_2 + Br \rightleftharpoons HBr + H$$

$$H + Br_2 \rightarrow HBr + Br.$$

Rate constants

In the vast majority of reaction sequences, i.e. mechanisms, the elementary reactions are bimolecular processes. The rate constant k for a bimolecular reaction

$$A + B \rightarrow \text{products}$$

can be defined by a rate equation

$$\frac{-d[A]}{dt} = k[A][B].$$

In other words, it is the rate with unit concentrations of the two reactants. If the time t is in seconds, and reactant concentrations are in mole l^{-1}, the dimensions of k are l mole^{-1} sec^{-1}. For a unimolecular reaction

$$A \rightarrow \text{products},$$

the dimensions of k are sec^{-1}, and for a termolecular process

$$A + B + C \rightarrow \text{products},$$

the dimensions of k are l^2 mole^{-2} sec^{-1}.

It is hoped, ultimately, that it will be possible to calculate rate constants by considering fundamental properties of individual atoms and molecules. Collision theory and transition-state theory, which are considered in later sections of this chapter, represent attempts, so far made, to relate measured rate constants with more fundamental processes.

Reactions in different phases

The simplest reactions to consider, are those in the gas phase, since they are not complicated by solvation effects. A few reactions have been studied in both the gas phase and in solution, but these rarely show much agreement. The decomposition of dinitrogen pentoxide

$$2N_2O_5 \rightarrow 4NO_2 + O_2,$$

is something of an exception, in that rates measured in eight different solvents agree to within a factor of two with those for the gas-phase reaction. Contrast the reaction of oxalyl chloride with water

$$(COCl)_2 + H_2O \rightarrow CO + CO_2 + 2HCl,$$

which is a complex chain reaction in the gas phase, but, in carbon tetrachloride, shows excellent second-order behaviour.

The ionic character of a substance is often emphasized in polar solvents. Thus, in aqueous solutions, hydrogen iodide behaves as a strong acid

$$HI \rightleftharpoons H^+ + I^-, (K \sim 10^7)$$

or, more precisely,

$$H_2O + HI \rightleftharpoons H_3O^+ + I^-,$$

and could hardly be expected to decompose, as in the gas phase

$$2HI \rightarrow H_2 + I_2.$$

For reactions in the gas phase, radicals are formed much more readily than ions, and ionic processes are practically non-existent below 800°C. Highly reactive ions can, however, be produced by electron bombardment in the ionization chamber of a mass spectrometer, and in a number of simple cases their reactions have been studied,

e.g. $$Ne^+ + H_2 \rightarrow HNe^+ + H.$$

For the many reactions in solution involving inorganic ions, water is by far the most convenient solvent. Acetic acid, methanol and ethanol have also been used, and liquid ammonia would no doubt be more popular were it easier to handle. Ethylene glycol-water, and other mixtures, have been used to study the effect of a variation in bulk dielectric constant.

In solution, ions interact with the solvent and other molecules present. Interactions are particularly strong for transition-metal ions which form a wide range of complexes. In aqueous solutions, hexaquo ions are generally formed; thus Cr^{3+} exists as the hexaquo ion $Cr(H_2O)_6^{3+}$. The affinity for water molecules is by no means exhausted by those in the inner-coordination sphere and there are further interactions with molecules at not very much greater distances. With non-transition metal ions, e.g. Na^+, electrostatic interactions with solvent dipoles are much weaker and the number of nearest neighbours is often difficult to establish. For a detailed discussion of the properties of metal ions in solution, the reader is referred to Hunt's *Metal Ions in Aqueous Solutions.*[1]

Stereochemistry of transition metal ions[2]

The number of groups in the inner-coordination sphere of a transition metal ion is generally either six (octahedral complexes) or four (tetrahedral and square-planar complexes). Most ions show variable coordination numbers, thus in aqueous solution the cobalt ion Co^{2+} has six water molecules as nearest neighbours, but at high chloride ion concentrations, tetrahedral $CoCl_4^{2-}$ is formed.

Although X-ray crystallographic techniques can be used to determine a coordination number in the solid phase, it cannot always be assumed that a complex has the same configuration in solution. Absorption spectra generally change appreciably with coordination number and provide one of the best methods of checking a particular configuration. Even so, it is not always easy to predict what the spectrum of an alternative configuration might be, or to say whether an observed shift is at all relevant. A good example to consider is that of the nickel(II) ion. While it is true that in the crystalline state Ni^{2+} often exists as $Ni(H_2O)_6^{2+}$, how do we know that a tetrahedral aquo-ion is not formed in aqueous solution? To answer this with any certainty, we must know what the spectrum of Ni^{2+} in a tetrahedral H_2O environment looks like. This in itself is difficult, but the spectrum of Ni^{2+} in a tetrahedral oxide environment can be obtained by

1. J. P. Hunt, *Metal Ions in Aqueous Solutions*, Benjamin, 1963.
2. H. Taube, *Progress in Stereochemistry* (Edited by de la Mare and Klyne), Vol. 3, p. 95, Butterworth's Scientific Publications, 1962.

dissolving small amounts of NiO in a ZnO lattice. Similarly, by dissolving NiO in MgO, the spectrum of Ni^{2+} in an octahedral oxide environment can be obtained. Since the spectrum of NiO in MgO resembles that of $Ni(H_2O)_6^{2+}$ in a crystal lattice and Ni^{2+} in aqueous solutions, it can be concluded that hexaquo ions are formed in solution.

The spectrum of $Fe(ClO_4)_3$ in perchloric acid solutions is very similar to that of $Fe(H_2O)_6^{3+}$ in an alum. If Fe^{3+} adopted a tetrahedral configuration in water, appreciable changes would be expected. The monochloro complex is also octahedral, $Fe(H_2O)_5Cl^{2+}$, but at higher chloride ion concentrations, tetrahedral $FeCl_4^-$ is formed and the spectrum resembles that of $KFeCl_4$. The stage at which the configuration actually changes can often be inferred from stepwise formation constants, where for successive reactions,

$$A + B \rightleftarrows AB$$

$$AB + B \rightleftarrows AB_2$$

stepwise formation constants may be defined by

$$K' = \frac{[AB]}{[A][B]} \text{ and } K'' = \frac{[AB_2]}{[AB][B]}.$$

Thus for the formation of $HgCl_4^{2-}$ such constants are found to be $K_1 = 10^{7 \cdot 15}$, $K_2 = 10^{6 \cdot 9}$, $K_3 = 10^{1 \cdot 0}$ and $K_4 = 10^{0 \cdot 7}$, and since K_1 and K_2 are very much bigger than K_3 and K_4, it can be inferred that the mono- and di-chloro complexes are linear, while the tri- and tetra-chloro complexes have tetrahedral configurations.

From a comparison of the spectra of VO^{2+} and $VO(acac)_2$ (the structure of which is known with certainty) the VO^{2+} aquo ion is believed to have a distorted octahedral structure $VO(H_2O)_5^{2+}$, in which four of the water molecules lie in a plane somewhat below that of the vanadium atom and on the side remote from the oxygen. Evidence that actinide ions such as the uranyl ion UO_2^{2+} are linear oxo-cations in solution as well as in the solid phase has been obtained by comparing visible infrared and Raman spectra.

Although such ions form a variety of complexes with other ions and donor molecules, few structures have as yet been determined. It is thought that four, five and six ligand groups can lie in the equatorial plane of, for example, the $O—U—O^{2+}$ ion.

Crystal-field theory: High-spin and low-spin complexes

Since in the chapters on solution reactions we shall be principally concerned with the reactions of octahedral metal ions (in particular the reactions of hexaquo metal ions), the discussion in this section will be limited to a consideration of such complexes. For a fuller account of crystal-field theory including that for tetrahedral and square-planar complexes, the reader should consult Orgel's *An Introduction to Transition-Metal Chemistry*.(3)

In a large number of complexes, the degree of metal–ligand orbital overlap is small, and crystal-field theory serves as an extremely useful first approximation. The theory is concerned with the electrostatic effect of the ligands, which are always either negatively charged or have a dipole with the negative end towards the metal. Thus, in an octahedral complex, the $d_{x^2-y^2}$ and d_{z^2} orbitals in the direction of the ligand groups have higher energies than the d_{xy}, d_{zy} and d_{zx} orbitals which point away from the ligands. For a regular octahedral environment, the five d-orbitals are split into two levels as shown in Fig. 1. Since calculations using crystal-field theory are generally comparative and do not have to take into account absolute energies, it can, for convenience, be assumed that the mean energy of the e_g and t_{2g} levels is zero when all the orbitals are equally occupied. The doubly degenerate e_g orbitals must, therefore, lie $\frac{3}{5}\Delta$ above, and the triply degenerate t_{2g} orbitals $\frac{2}{5}\Delta$ below the energy of the unperturbed d-orbitals.

The colours of transition metal ions are, in the majority of cases, due to electron transitions between the t_{2g} and e_g levels. The energy of separation Δ can therefore be obtained from

3. L. E. Orgel, *An Introduction to Transition-Metal Chemistry*, Methuen, 1960.

absorption spectra. For example $Ti(H_2O)_6^{3+}$ has an absorption peak in the visible at 20,000 cm^{-1}, and since 350 cm^{-1} is equivalent to one kilocalorie, the energy of separation Δ is $\sim$57 kcal $mole^{-1}$.

While there is only one possible ground state for ions with one, two or three d electrons (i.e. t_{2g}^1, t_{2g}^2 and t_{2g}^3), and for ions with eight or nine d electrons (i.e. $t_{2g}^6e_g^2$ and $t_{2g}^6e_g^3$), with d^4, d^5, d^6 and d^7 ions, there are two possibilities. The configuration with the

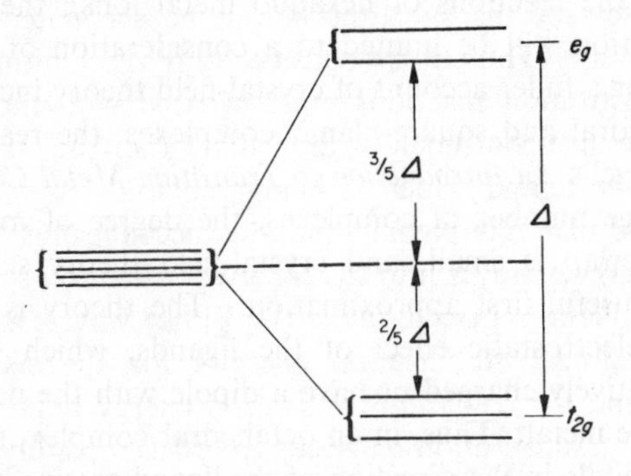

FIG. 1. The splitting of a set of d-orbitals in a symmetrical octahedral crystal field.

maximum possible number of unpaired electrons is known as the high-spin state, and that with a minimum as the low-spin state. With, for example, a d^5 ion, the two possibilities are t_{2g}^5 and $t_{2g}^3e_g^2$. Which of these two represents the ground state for any one set of ligands can only be answered by comparing Δ and P values, where P is the energy required to cause electron pairing of two electrons in the same orbital. For a t_{2g}^5 ion, the five electrons in the t_{2g} level represent an energy stabilization of $5 \times -\frac{2}{5}\Delta$, less P for each pair of electrons occupying the same orbital. For the spin-free $t_{2g}^3e_g^2$ ion, on the other hand, there is no stabilization. The difference in energy of the two forms is, therefore, $-2\Delta + 2P$ and the existence of the spin-paired t_{2g}^5 form requires Δ to be greater than P. If Δ is less than P, then the complex will be high-spin.

A feature of the first transition series is the existence of a large number of complexes of $Cr^{III}(t_{2g}^3)$ and spin-paired $Co^{III}(t_{2g}^6)$. In both series of complexes the ligands are replaced only very slowly (half-lives of the order of 1–2 days) and they are said to be non-labile or inert, e.g.

$$Cr(H_2O)_6^{3+} + H_2{}^{18}O \xrightarrow{t_{\frac{1}{2}} \sim 40hr} Cr(H_2O)_5(H_2{}^{18}O)^{3+} + H_2O.$$

The slowness of the substitution can be accounted for in the following way. Thus, if we consider only the limiting cases, all ligand substitution reactions must proceed with the formation of either a five-coordinate intermediate (i.e. the mechanism is S_N1 and the ligand which is being replaced leaves first), or a seven-coordinate intermediate (i.e. the mechanism is S_N2 and the incoming group moves into the inner-coordination sphere before the group being replaced has left). Either way, because of the high crystal-field stabilization energies for the six-coordinate $Cr^{III}(t_{2g}^3)$ and $Co^{III}(t_{2g}^6)$ ions, activation energy requirements are too large for substitution reactions to proceed readily.[4]

As a general rule, high-spin complexes are more labile than low-spin complexes (according to one definition, an ion is labile if ligand substitution reactions are complete within the time of mixing, say, 1 min, at room temperature, and with 0·1 M solutions). While Co^{III} complexes are invariably spin-paired, with Fe^{II} (which is also d^6), only the *o*-phenanthroline, α, α′-bipyridine and CN^- ligands have sufficiently strong crystal fields to cause electron pairing. Thus, $Fe(H_2O)_6^{2+}$ is high-spin and extremely labile while $Fe(phen)_3^{2+}$, $Fe(bipy)_3^{2+}$ and $Fe(CN)_6^{4-}$ are spin-paired and inert. Successive stability constants for the formation of $Fe(phen)_3^{2+}$ from $Fe(H_2O)_6^{2+}$ are $K_1 = 10^{5\cdot85}$, $K_2 = 10^{5\cdot25}$ and $K_3 = 10^{10}$. These suggest that spin-pairing occurs in going from $Fe(phen)_2(H_2O)_2^{2+}$ to $Fe(phen)_3^{2+}$.

4. F. Basolo and R. G. Pearson, *Mechanism of Inorganic Reactions*, p. 108, Wiley, 1958.

The effect of anions

Oxidation–reduction reactions involving hexaquo metal ions are generally very dependent on the type of anion present. A labile ion such as hexaquo Fe^{3+} will, for example, readily substitute a Cl^- for an H_2O in its inner sphere and, since Fe^{3+} and

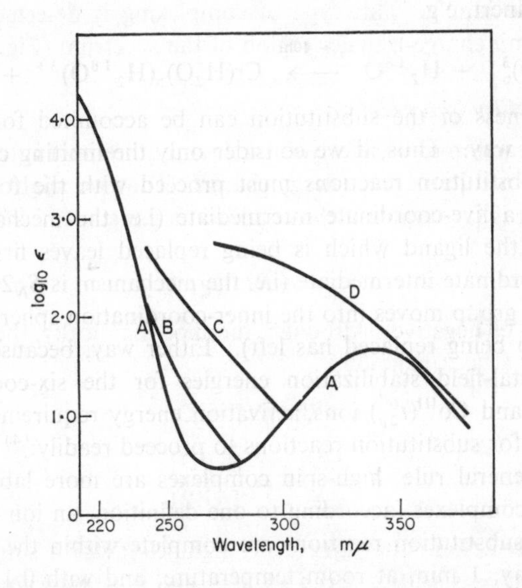

FIG. 2. The effect of outer-sphere (or ion-pair) complexing on the absorption spectrum of $Co(NH_3)_6{}^{3+}$ (A). Curve B is with 0·9 M Cl^-, C with 0·9 M Br^- and D with 0·9 M I^-. [Reproduced, with permission, from M. G. EVANS and G. H. NANCOLLAS, *Trans. Faraday Soc.* **49**, 363 (1953).]

$FeCl^{2+}$ react at different rates, the overall rate will be affected. Of the common anions, perchlorate has least tendency to complex, and is therefore the most widely used. It is more difficult to establish whether the perchlorate ion itself forms complexes. Spectral evidence has been obtained for such complexes with Ce^{3+}, Hg^{2+} and Fe^{3+}, but that of Fe^{3+}, at least, is believed to be of the outer-sphere type.[5] Perchlorate ions slowly oxidize V^{2+}, Ti^{3+} and V^{3+}, but are otherwise remarkably stable to a wide range of cations.

Anion interactions are not always confined to the inner-coordination sphere, and spectral evidence has been obtained for ion-pair or outer-sphere $Co(NH_3)_6^{3+}$, X^- complexes where X^- is a halide ion.[6] This type of complexing is detected in the 250–350 mμ charge-transfer region of the spectrum (Fig. 2).

The feasibility of a reaction

For a reaction to be effective, the first requirement is that there should be a suitable free energy change. Thus we might say that a reaction is favourable if the equilibrium constant K is >1, and since

$$\Delta G^\circ = -RT \log_e K,$$

ΔG° must be negative. Although, thermodynamically, there must always be a back reaction,

$$A + B \rightleftharpoons C + D,$$

where
$$K = \frac{k_{AB}}{k_{DC}} = \frac{[C][D]}{[A][B]},$$

the back reaction is not kinetically significant (for a single-stage reaction) if K is $>10^2$. Even so, if K and k_{AB} are known, k_{CD} for the back reaction can always be determined. The free energy change for a gas reaction can be calculated from standard free energies (see Appendix 1), and for reactions in aqueous solution from oxidation–reduction potentials (see Appendix 2).

Although the free energy change will indicate the feasibility of a reaction, it generally gives no information as to the rate. The free energy changes for the reaction of hydrogen with oxygen, and hydrogen with fluorine, are both extremely favourable, yet a

---

5. K. W. Sykes, *J. Chem. Soc.* 2473 (1959).

6. M. G. Evans and G. H. Nancollas, *Trans. Faraday Soc.* **49**, 363 (1953); E. L. King, J. H. Espenson and R. E. Visco, *J. Phys. Chem.* **63**, 755 (1959).

stoicheiometric mixture of hydrogen and oxygen can be kept indefinitely at a room temperature, while the corresponding hydrogen–fluorine mixture is explosive under identical conditions. The second requirement, then, for a reaction to be feasible at a given temperature, is that there should be a suitable mechanism with sufficiently low energy requirements. For a number of related reactions Linear Free Energy Relationships ($\Delta G^{\ddagger}$ against ΔG^{0}) have been observed.

The reaction products and stoicheiometry of a reaction

Preliminary experiments are generally concerned with identifying the products and determining the stoicheiometry of a reaction. While, for many reactions, these may seem obvious, a check to see whether other possible products are completely absent is often worthwhile. In the decomposition of nitrous oxide, for example, the reaction is predominantly

$$2N_2O \rightarrow 2N_2 + O_2,$$

but small amounts of nitric oxide are also formed,

$$2N_2O \rightarrow N_2 + 2NO.$$

Since, for each path, two volumes of nitrous oxide give three volumes of product, the measurement of pressure changes alone does not allow a full kinetic treatment, and in this example a separate determination of nitric oxide is also necessary.

In the gas-phase decomposition of hydrogen peroxide, on the other hand, that water and oxygen are products and that no hydrogen is formed, is clearly evidence for a first step involving the breaking of an O—O bond. If hydrogen atoms were formed at any stage, then hydrogen would almost certainly be one of the products.

For reactions in solution, side reactions with the solvent can be important. Thus in the reaction between thallium(I) and cobalt(III),

$$Tl^{I} + 2Co^{III} \rightarrow Tl^{III} + 2Co^{II},$$

Co^{III} oxidizes water, and only by a careful choice of reactant concentrations can the latter be reduced to less than 10% of the total reaction. Other metal ions, notably those of chromium(II) and vanadium(II), react rapidly with atmospheric oxygen, and reactions have to be carried out under an inert atmosphere.

When the initial products undergo further reaction the approach can often be simplified by the separate study of subsequent reactions. In the reaction of vanadium(III) with cobalt(III), for example, the initial step

$$V^{III} + Co^{III} \rightarrow V^{IV} + Co^{II},$$

is followed by the further oxidation of V^{IV}

$$V^{IV} + Co^{III} \rightarrow V^{V} + Co^{II},$$

and an interpretation of the kinetics is easier once this latter reaction has been studied. Similarly, in the decomposition of dinitrogen pentoxide

$$2N_2O_5 \rightarrow 4NO_2 + O_2,$$

existing (literature) data is used to allow for the rapid subsequent equilibrium

$$2NO_2 \rightleftharpoons N_2O_4.$$

The extent of this reaction under a given set of conditions obviously effects either absorptiometric measurements of $[NO_2]$, or manometric measurements of the progress of the reaction.

Temperature dependence of rate constants

Chemical reactions are generally very sensitive to temperature and, as a rough sort of guide, the rate will often double for a 10°C rise in temperature. Arrhenius (1889) found the variation of a rate constant k with temperature could be expressed by the equation

$$\log_{10} k = \log_{10} A - \frac{E}{2{\cdot}303\, RT}$$

or

$$k = A\mathrm{e}^{-E/RT}.$$

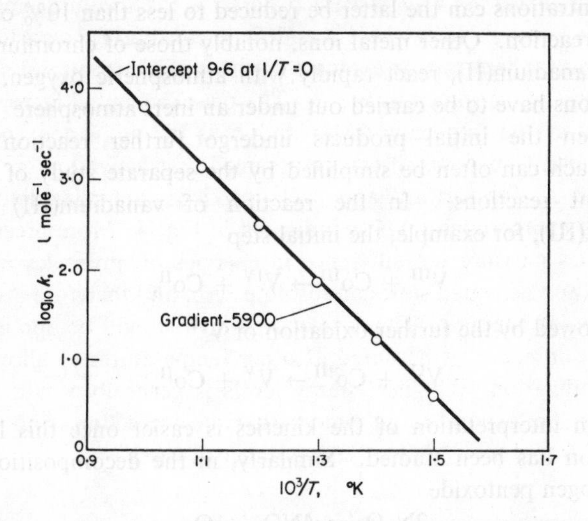

FIG. 3. Arrhenius plot for the bimolecular reaction $2NO_2 \rightarrow 2NO + O_2$. The intercept and slope are equal to $\log_{10} A$ (l mole^{-1} sec^{-1}) and $-E/2{\cdot}303\ R$ (cal) respectively.

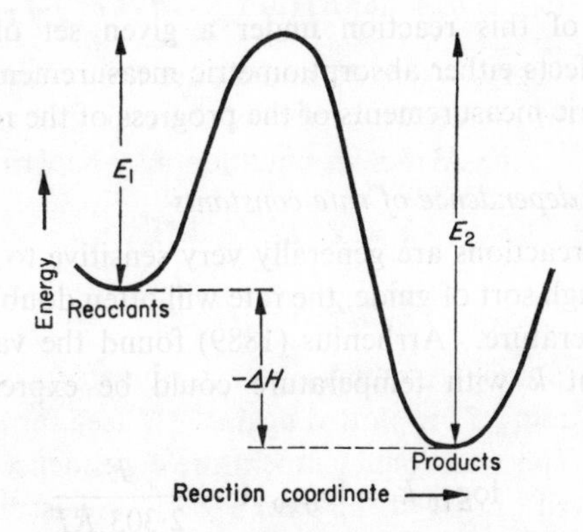

FIG. 4. Diagram indicating energy requirements for a chemical reaction.

Thus, a plot of log k against $1/T$ (where T is the absolute temperature) is a straight line, where A and E are constant for a particular reaction. These can be evaluated from the intercept and slope, respectively, as shown in Fig. 3. The pre-exponential factor A is known as the frequency factor and E as the activation energy. For present purposes, it can be assumed that activation energies E_1 and E_2 for forward and back reactions, are related to the overall exothermicity ΔH as indicated in Fig. 4. More precisely, E values measure the difference in energies of molecules in the initial and activated states at absolute zero, and therefore involve zero point energies. The correction which should be applied to activation energies measured at other temperatures to allow for the population of higher energy levels is generally small. For reactions at around room temperature the correction is no more than 0·6 kcal mole^{-1} in a large number of cases.

The collision theory of bimolecular reactions

According to simple collision theory, the pre-exponential A factor should be equivalent to the bimolecular collision frequency Z. The latter can be calculated from the expression

$$Z = (\sigma_A + \sigma_B)^2\,[8\pi RT(M_A + M_B)/M_A M_B]^{1/2}$$

from kinetic theory, where σ_A and σ_B are the collision radii and M_A and M_B the molecular weights of the two reactants A and B. Since $Z \propto T^{1/2}$, the Arrhenius equation may be written

$$k = Z'\,T^{1/2}e^{-E/RT},$$

where

$$Z = Z'\,T^{1/2},$$

and a more precise Arrhenius plot is therefore one in which $(\log_{10} k - \frac{1}{2}\log_{10} T)$ is plotted against $1/T$ (see for example Fig. 19 on p. 87). This additional temperature dependence is, however, small compared to that of the term $e^{-E/RT}$, and is often ignored. The term $e^{-E/RT}$ represents the fraction of collisions which have sufficient activation energy E for reaction to take place.

When Z and A are compared, there is reasonable agreement only for those reactions involving simple molecules. In a very large number of reactions, they differ by many powers of ten, and so A is often divided into a true collision number Z multiplied by a steric or probability factor P. Thus

$$k = PZ\,e^{-E/RT}.$$

Experimental values of P may be as small as 10^{-8} for reactions involving complex organic molecules. Such values are attributed to steric effects, since the orientation of molecules to each other at the time of collision will obviously effect their chances of reacting.

Although collision theory provides a reasonably successful model for bimolecular reactions, it is less successful in accounting for the rates of unimolecular and termolecular processes. It is also difficult to apply to solution reactions, since, in the presence of solvent, the expression for the collision frequency from gas kinetic theory is, strictly speaking, no longer applicable.

Transition-state theory

For solution reactions at least collision theory has been largely supplanted by transition-state theory. For a reaction between A and B a transition state $AB^{\ddagger}$ is considered:

$$A + B \rightarrow AB^{\ddagger}\ \text{products.}$$

If equilibrium conditions are assumed then the concentration of $AB^{\ddagger}$ is given by

$$[AB^{\ddagger}] = K^{\ddagger}\,[A][B],$$

provided activities can be identified with concentrations. The intermediate $AB^{\ddagger}$ is treated as a normal molecule, except that one of its vibrations is now equivalent to a translational degree of freedom along the direction which leads to the formation of reaction products. The product ν of the frequency of this vibration and a vibrational component of the partition function of the complex, is given by

$$h\nu = kT,$$

where k is Boltzmann's and h is Planck's constant. The rate of reaction is therefore

$$-d[\mathrm{A}]/dt = \nu[\mathrm{AB}^{\ddagger}] = (kT/h)\,[\mathrm{AB}^{\ddagger}].$$

But the experimental rate constant is defined by the rate equation

$$-d[\mathrm{A}]/dt = k[\mathrm{A}][\mathrm{B}],$$

and so

$$k = \frac{kT}{h}\frac{[\mathrm{AB}^{\ddagger}]}{[\mathrm{A}][\mathrm{B}]} = \frac{kT}{h}K^{\ddagger}.$$

Now $K^{\ddagger}$, like any other equilibrium constant, is related to the free energy of formation of the activated complex, $\Delta G^{\ddagger}$, by the thermodynamic expression

$$\Delta G^{\ddagger} = -RT\log_{e}K^{\ddagger},$$

whence

$$k = (kT/h)\,\mathrm{e}^{-\Delta G^{\ddagger}/RT}.$$

Since $\Delta G^{\ddagger}$ is related to the heat of activation $\Delta H^{\ddagger}$ and entropy of activation $\Delta S^{\ddagger}$ by the relationship

$$\Delta G^{\ddagger} = \Delta H^{\ddagger} - T\Delta S^{\ddagger},$$

the above expression can be written

$$k = (kT/h)\,\mathrm{e}^{\Delta S^{\ddagger}/R}\,\mathrm{e}^{-\Delta H^{\ddagger}/RT}.$$

For experiments at temperatures of around 300°K, $kT/h = 6{\cdot}13 \times 10^{12}$ sec^{-1}.

It is useful to be able to express this equation in a form which involves the experimental activation energy E in place of the heat of activation $\Delta H^{\ddagger}$. For unimolecular reactions and all reactions in solution it can be shown that[7]

$$E = \Delta H^{\ddagger} + RT,$$

---

7. K. J. Laidler, *Reaction Kinetics*, Vol. 1, p. 86, Pergamon Press, London, 1963.

so that k may be expressed

$$k = (kT/h)\,e^{\Delta S^{\ddagger}/R}\,e^{-(E-RT)/RT},$$

or in the form

$$k = (e\,kT/h)\,e^{\Delta S^{\ddagger}/R}\,e^{-E/RT}.$$

For bimolecular reactions in the gas phase, the relationship between E and $\Delta H^{\ddagger}$ has to be modified however, and

$$E = \Delta H^{\ddagger} + 2RT,$$

so that

$$k = (e^{2}kT/h)\,e^{\Delta S^{\ddagger}/R}\,e^{-E/RT}.$$

The above expressions may be compared to the collision theory expression

$$k = PZ\,e^{-E/RT}.$$

Thus, for unimolecular and solution reactions, it follows that

$$PZ = (ekT/h)e^{\Delta S^{\ddagger}/R}.$$

In principle, $\Delta S^{\ddagger}$ can be calculated if vibration frequencies, bond-length, etc., are known for the intermediate $AB^{\ddagger}$. Little is known of such transition complexes, however, and these quantities are not always easy to estimate. The main advantage of the theory is that it can be applied equally well to any reaction system, whether in solution or in the gas phase.

In the chapters to follow, activation energies E from collision theory are used for gas-phase reactions, and heats of activation $\Delta H^{\ddagger}$ from transition state theory for solution reactions. The latter are often inaccurately described as activation energies; since $E = \Delta H^{\ddagger} + RT$, E and $\Delta H^{\ddagger}$ differ by only 0·6 kcal mole^{-1} at room temperature.

The effect of ionic strength

Charged particles in solution are affected not only by their nearest neighbours, which, in the case of transition metal ions, are coordinated groups, but also by other ions at greater distances. If the non-nearest neighbours of an ion are changed, then the potential energy and activity coefficient of that ion will be affected. A positive ion will, for instance, occur in a negatively charged environment, the environment changing as the ionic strength of the solution is changed. The ionic strength μ is given by the expression

$$\mu = \tfrac{1}{2}\Sigma mz^2$$

where m is the molarity of any one ion and z its charge.

Only in dilute solution ($\mu < 0{\cdot}01$ M), can the effect of ionic strength variations be predicted. The effective concentration or activity of an ion is the product of its concentration c and its activity coefficient f,

$$a = cf.$$

In dilute solutions f is < 1, and at infinite dilution it approaches unity. From transition-state theory, the rate of the reaction

$$A + B \rightleftharpoons AB^{\ddagger} \rightarrow \text{products}$$

is given by

$$k = \frac{kT}{h}\frac{[AB^{\ddagger}]}{[A][B]}$$

as shown above. If now we consider the case in which A and B are ions,

$$K^{\ddagger} = \frac{a_{AB^{\ddagger}}}{a_A a_B} = \frac{[AB^{\ddagger}]}{[A][B]}\frac{f_{AB}}{f_A f_B},$$

then

$$k = \frac{kT}{h}K^{\ddagger}\frac{f_A f_B}{f_{AB^{\ddagger}}} = \frac{k_0 f_A f_B}{f_{AB^{\ddagger}}},$$

where k_0 is the rate constant at infinite dilution when all the activity coefficients are unity. In dilute solutions with the ionic strength below 0·01 for a 1 : 1 electrolyte and below 0·001 M for some higher charged ions, the Debye–Hückel limiting law is applicable,

$$\log_{10} f = -Az^2\sqrt{\mu},$$

where A is a constant which is inversely proportional to the dielectric constant D. It follows, therefore, that

$$\log_{10}(f_A f_B / f_{AB^\ddagger}) = -A\sqrt{\mu}\,(z_A^2 + z_B^2 - z_{AB^\ddagger}^2),$$

but since

$$z_{AB^\ddagger} = z_A + z_B,$$

$$\log_{10}(f_A f_B / f_{AB^\ddagger}) = 2Az_A z_B\sqrt{\mu}.$$

In aqueous solutions at 25°, $A = 0{\cdot}5$, and substituting for $\log_{10}(f_A f_B/f_{AB^\ddagger})$ in the expression

$$\log_{10} k = \log_{10} k_0 + \log_{10}(f_A f_B / f_{AB^\ddagger})$$

we get

$$\log_{10} k = \log_{10} k_0 + z_A z_B \sqrt{\mu}$$

or

$$\log_{10}(k/k_0) = z_A z_B \sqrt{\mu}.$$

The latter is known as the Brønsted–Bjerrum equation. It is illustrated in Fig. 5, where $\log_{10}(k/k_0)$ is plotted against $\mu^{1/2}$; in each case, the slope is equal to the charge product $z_A z_B$[8]. For reactions between ions of the same charge type, k increases with increasing ionic strength, and for reactions between ions of different charge, k decreases with increasing ionic strength.

Reactions are generally performed in a medium at constant ionic strength, $\mu = 3{\cdot}0$ M, being the highest ionic strength commonly used. Reactant concentrations generally range from 10^{-1} to 10^{-5} M, depending on the speed of the reaction and the sensitivity of the method used to follow its progress. Fairly high ionic strengths of 0·5 to 3·0 M are preferable, so that effects resulting from a variation in reactant concentrations are negligible. If, in the reaction of two metal ions of different charge type, the ionic strength of the perchlorate medium is small and just sufficient to permit variations in reactant concentrations, then, although the ionic strength might be constant, the concentration

---

8. J. Barrett and J. H. Baxendale, *Trans. Faraday Soc.* **52**, 210 (1956).

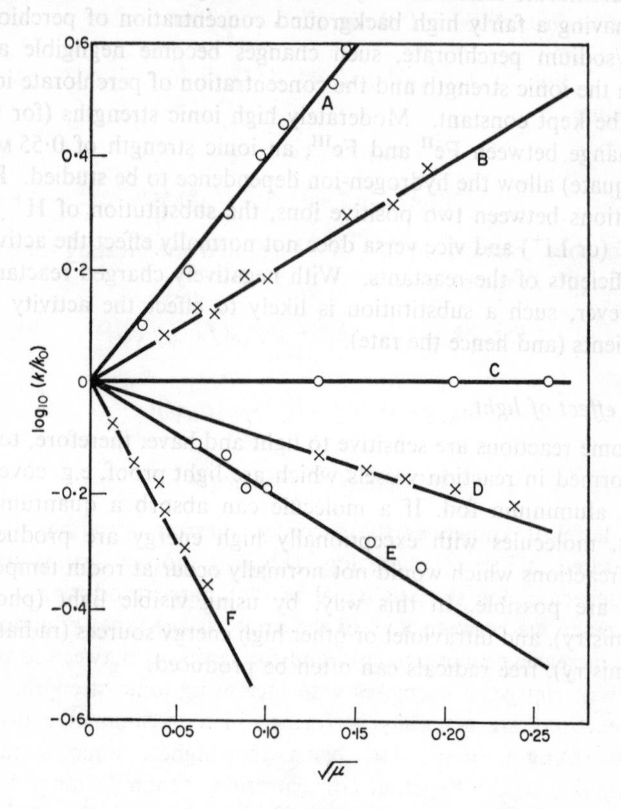

FIG. 5. The dependence of reaction rates on ionic strength. The reactions are:

A. $Co(NH_3)_5Br^{2+} + Hg^{2+} + H_2O \rightarrow Co(NH_3)_5H_2O^{3+} + HgBr^+$
B. $S_2O_8^{2-} + 3I^- \rightarrow 2SO_4^{2-} + I_3^-$
C. $Cr(urea)_6^{3+} + 6H_2O \rightarrow Cr(H_2O)_6^{3+} + 6urea$
D. $H_2O_2 + 2H^+ + 2Br^- \rightarrow 2H_2O + Br_2$
E. $Co(NH_3)_5Br^{2+} + OH^- \rightarrow Co(NH_3)_5OH^{2+} + Br^-$
F. $Fe^{2+} + Co(C_2O_4)_3^{3-} \rightarrow Fe^{3+} + Co(C_2O_4)_3^{4-}$.

of perchlorate ions will vary from one experiment to the next. By having a fairly high background concentration of perchloric acid/sodium perchlorate, such changes become negligible and both the ionic strength and the concentration of perchlorate ions can be kept constant. Moderately high ionic strengths (for the exchange between Fe^{II} and Fe^{III}, an ionic strength of 0·55 M is adequate) allow the hydrogen-ion dependence to be studied. For reactions between two positive ions, the substitution of H^+ for Na^+ (or Li^+) and vice versa does not normally effect the activity coefficients of the reactants. With negatively charged reactants, however, such a substitution is likely to effect the activity coefficients (and hence the rate).

The effect of light

Some reactions are sensitive to light and have, therefore, to be performed in reaction vessels which are light proof, e.g. covered with aluminium foil. If a molecule can absorb a quantum of light, molecules with exceptionally high energy are produced, and reactions which would not normally occur at room temperature are possible. In this way, by using visible light (photochemistry), and ultraviolet or other high energy sources (radiation chemistry), free radicals can often be produced.

CHAPTER 2

EXPERIMENTAL TECHNIQUES

A VARIETY of physical and chemical methods have been used to follow the progress of chemical reactions. These fall into two groups, conventional techniques being used for reactions with half-lives, $t_{\frac{1}{2}}$, of more than 30 sec, and fast reaction techniques for reactions with half-lives down to 10^{-9} sec. In bimolecular reactions with initial reactant concentrations both c, the rate constant is given by $1/ct_{\frac{1}{2}}$, where, for solution reactions, c is generally of the order 10^{-2}–10^{-3} M. When physical methods are used, it is first necessary to check that the change in physical property is proportional to the change in concentration, i.e. that Beer's Law holds when spectrophotometric measurements are made. As a general rule, absorptiometric and, for gas-phase reactions, manometric methods are used whenever possible.

A. CONVENTIONAL

Volumetric methods

Standard volumetric procedures have been used for a number of oxidation–reduction reactions between metal ions, when absorptiometric methods are not applicable. The reaction of iron(II) with thallium(III) can, for example, be quenched by adding excess cerium(IV) solution (which oxidizes the unreacted Fe^{II}) and titrating the excess of Ce^{IV} against standard Fe^{II}. Similarly, in the reaction of mercury(I) with manganese(III) at 50°C samples of the reaction solution can be quenched by cooling to room temperature, and the Mn^{III} then estimated by using 10^{-3} N Fe^{II}, with ferroin as indicator.

Reactions in which halide ions are liberated can be followed by titrating with silver nitrate,

e.g. $Co(en)_2\,Cl_2^+ + H_2O \rightarrow Co(en)_2\,Cl\,H_2O^{2+} + Cl^-$.

In this particular case, reaction solutions are sufficiently coloured to obscure the end-point, and titrations are performed potentiometrically, using a silver/silver chloride electrode.

In the gas-phase reaction between hydrogen and iodine over the temperature range 283–508°C, reaction mixtures are first quenched by rapid cooling. The seal on the reaction vessel is then broken under dilute sodium hydroxide solution, and the iodide and iodine concentrations determined by standard procedures. The volume of hydrogen gas remaining can be measured.

Pressure measurements

For gas-phase reactions, some experience of glass blowing and high-vacuum techniques is generally required. Reactions in which there is a decrease in the number of molecules, e.g.

$$CO + Cl_2 \rightarrow COCl_2,$$
$$2NO + O_2 \rightarrow 2NO_2,$$
$$2N_2O_5 \rightarrow 4NO_2 + O_2,$$

are conveniently followed by measuring the pressure changes with time (Fig. 6). Pressures down to 10^{-3} mm are obtained using a rotational oil pump; if, in addition, a mercury diffusion pump is used, pressures of 10^{-6} mm Hg are possible. More recently, pressures in the region 10^{-9}–10^{-10} mm have been obtained with mercury cut-offs or mercury taps and an adequate take-out procedure. Pressures in mm Hg can readily be converted into concentration terms (mole l^{-1}), since 1 mole of a gas at NTP occupies 22·4 litres. For reactions in which there is no pressure change, e.g.

$$H_2 + Cl_2 \rightarrow 2HCl,$$

a modified procedure is often possible. In this particular case, the hydrogen chloride and chlorine can be condensed by means of cold traps, and the amount of residual hydrogen gas measured.

The homogeneity of gas reactions in flow as well as static methods should always be checked. This is most readily done by changing the surface-area to volume ratio of the reactor, e.g. by filling the reactor with glass beads. If the rate increases with a greater surface-area to volume ratio, then the reaction is, at

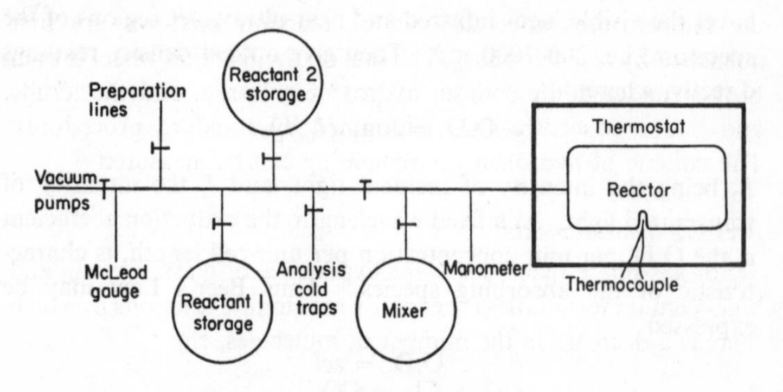

FIG. 6. Diagram of apparatus used for studying gas reactions by measuring pressure changes.

least in part, heterogeneous. To eliminate or reduce the heterogeneous reaction, the material of the vessel might be changed, e.g. quartz for glass, or the surface might be coated with some suitable material, e.g. potassium chloride or paraffin wax. Heterogeneous effects are generally reduced by working at higher temperatures. In the decomposition of hydrogen peroxide, the Arrhenius plot shows a fairly abrupt change at about 400°C corresponding to the change from essentially heterogeneous to homogeneous reaction.[1] Reaction vessels may be kept at a steady temperature using electrically heated furnaces or, over the

1. D. E. HOARE, J. B. PROTHEROE and A. D. WALSH, *Trans. Faraday Soc.* **55**, 548 (1959).

50–300°C region, by immersion in boiling organic liquids. If the reactant gases are corrosive and attack mercury, a Bourdon pressure gauge may be used.

Absorptiometric methods

With the availability of commercial spectrophotometers complete with thermostated cell housings, absorptiometric methods are now widely used for solution reactions. These instruments cover the visible, near infrared and near ultraviolet regions of the spectrum, i.e. 200–1000 mμ. They give optical density readings directly, where

$$\text{O.D.} = \log_{10}(I_0/I_t),$$

I_0 being the intensity of incident light, and I_t the intensity of transmitted light. At a fixed wavelength, the extinction coefficient ε, the O.D. per unit concentration per unit-cell length, is characteristic of the absorbing species. Thus Beer's Law may be expressed

$$\text{O.D.} = \varepsilon c l$$

where c is the concentration of absorbent and l the cell length. Beer's Law can be tested by making up a series of solutions of varying dilution and plotting O.D. values against concentration. If the law holds over the relevant range of concentrations, that is, if there is no association or dissociation of the absorbing species, then optical density readings can be used directly, as a measure of concentration.

Some typical spectra are shown in Fig. 7. These are relevant to the reaction

$$\text{Co}^{\text{III}} + \text{V}^{\text{IV}} \rightarrow \text{Co}^{\text{II}} + \text{V}^{\text{V}}$$

which has been studied by following O.D. changes at 400 mμ. Measurements at 600 mμ are less satisfactory since cobalt(II) and vanadium(IV) also absorb at this wavelength. Vanadium(V) ions absorb strongly in the near ultraviolet. In a number of cases in which one of the reactants has an exceptionally high absorption, it is possible to work with small reactant concentrations, and fairly fast rate constants can be measured in the normal way. Thus in the reaction

$$Co(C_2O_4)_3^{3-} + Fe^{2+} \rightarrow Co^{2+} + 3C_2O_4^{2-} + Fe^{3+}$$

the reactant $Co(C_2O_4)_3^{3-}$ has such a high absorption coefficient that reactant concentrations of the order 10^{-5} M can be used.

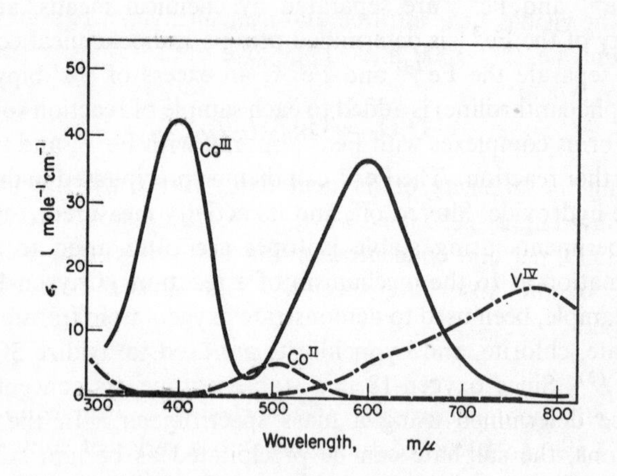

FIG. 7. Absorption spectra for cobalt(III) and vanadium(IV) in aqueous perchloric acid solutions.

Absorptiometric measurements in the visible can also be used to follow gas reactions. They have proved particularly useful in studying the recombination of nitrogen dioxide and halogen atoms,

$$NO_2 + NO_2 + M \rightarrow N_2O_4 + M$$

$$I + I + M \rightarrow I_2 + M,$$

where M is a third-body. Such reactions are fast, and electronic devices have to be used to record photographically the changes in light absorption.

The use of tracers

Electron-transfer reactions of the type

$$Fe^{2+} + {}^{*}Fe^{3+} \rightarrow Fe^{3+} + {}^{*}Fe^{2+}$$

can be studied by using radioactive isotopes, e.g. ^{59}Fe, to label one of the reactants.[2] The amount of tracer used depends on its activity, thus, in this particular instance, less than 1% of $^{59}Fe^{3+}$ is required to label the Fe^{3+}. To follow such reactions, the Fe^{2+} and Fe^{3+} are separated by chemical means, and the activity of the Fe^{3+} is determined using a radiochemical counter.

To separate the Fe^{2+} and Fe^{3+}, an excess of α,α′-bipyridine (or *o*-phenanthroline) is added to each sample of reaction solution. This forms complexes with Fe^{2+}, but not with Fe^{3+}, and there is no further reaction. The Fe^{3+} can then be precipitated in the form of the hydroxide, filtered off, and its activity measured.

Experiments using stable isotopes are often used to obtain information as to the mechanism of a reaction. Oxygen-18 has, for example, been used to demonstrate oxygen-atom transfer when chlorate, chlorite, and hypochlorite are used to oxidize SO_3^{2-} to SO_4^{2-}.[3] Since oxygen-18 is a stable isotope, its concentration can be determined using a mass spectrometer. In the above reactions, the sulphate can be precipitated as barium sulphate, filtered off and heated with carbon, and the resulting sample of carbon dioxide fed into the mass spectrometer.

Electrometric methods

A fairly wide range of rate constants (including some for fast reactions) can often be determined by following changes in electrical properties. The rate of the $Fe^{3+} + F^-$ complexing reaction has, for example, been measured by observing changes in the Fe^{2+}/Fe^{3+} half-cell potential, as fluoride ions form a complex with the Fe^{3+}.[4] The concentrations of Fe^{2+}, Fe^{3+} and F^- used, were of the order 10^{-4} l^{-1} mole.

---

2. J. Silverman and R. W. Dodson, *J. Phys. Chem.* **56**, 846 (1952).
3. J. Halperin and H. Taube, *J. Am. Chem. Soc.* **74**, 375 (1952).
4. W. MacF. Smith, *Proc. Chem. Soc.* 207 (1957).

In other systems, e.g.

$$Co(en)_2\,Cl_2^+ + H_2O \rightarrow Co(en)_2\,Cl(H_2O)^{2+} + Cl^-,$$

the conductivity of the solution changes as the reaction proceeds. These can therefore be followed by measuring changes in resistivity of the reaction solution, since the latter is determined by ionic conductances.

Rates can also be determined by recording pH changes. An accurate pH-meter reading to 0·002 of a pH unit has been used to follow the reaction

$$Co(NH_3)_5CO_3^+ + H_3O^+ \rightarrow Co(NH_3)_5H_2O + HCO_3^-,$$

the pH changing by 0·15 units, in a typical experiment.

Polarimetric methods

Polarimetric measurements can be used to study rates when the reactants are optically active. Thus the rate of the Co^{II}(PDTA)–Co^{III}(PDTA) electron exchange (where PDTA is the hexadentate ligand propylenediaminetetraacetate) can be determined by the reaction of optically active forms.[5] The reaction is essentially

$$dextro\text{-}Co^{II} + laevo\text{-}Co^{III} \rightarrow dextro\text{-}Co^{III} + laevo\text{-}Co^{II}.$$

Other systems are not always as readily interpreted, since the individual reactants sometimes undergo separate racemization. In the exchange between Co^{II}(EDTA) and Co^{III}(EDTA), for example, the Co^{II} complex undergoes instant racemization, i.e.

$$dextro\text{-}Co^{II} \overset{\text{fast}}{\rightleftharpoons} laevo\text{-}Co^{II}.$$

Other methods

Solution reactions which are accompanied by a volume change can be studied by a dilatometric method. In a dilatometer, the reaction solution extends into a piece of capillary tubing attached

5. Y. H. Im and D. H. Busch, *J. Am. Chem. Soc.* **83,** 3362 (1961).

to the reaction vessel, and the height of the solution is measured using a travelling microscope. These and viscosity measurements, using standard viscometers, are particularly useful for reactions involving polymerization.

Thermal conductivity measurements have been used to follow the progress of hydrogen reactions

$$D_2 + H_2 \rightarrow 2HD,$$

and

$$\text{para-}H_2 \rightarrow \text{ortho-}H_2.$$

The method is dependent on such molecules having different conductivities, thus H_2 is some 6% more effective than D_2 in conducting heat. The constitution of a gas mixture can be inferred from the temperature of a heated platinum filament inside a vessel containing a sample of the gas. As the reaction proceeds, the temperature, and therefore the resistance of the filament change, and with suitable calibration (using samples of known composition), the composition of a gas sample can be determined by measuring the resistance of the filament.

B. FAST REACTIONS

Flow methods for gas reactions

Flow methods generally give much less precise data for gas-phase reactions, and are used only when static systems are found to be inadequate. This might be because the reaction is unduly fast over a range of temperatures which are of particular interest, or because the system is unduly complex and a variety of secondary reactions might follow the initial step.

In the apparatus in Fig. 8 for reactions at elevated temperatures, the pre-heated gases are led into the reaction vessel through tubes which are arranged tangentially to it for efficient mixing. The time the reactants spend in the reaction vessel is obtained from flow rates (flow meters calibrated in $cm^3\ sec^{-1}$ are used for each of the reactant gas streams) and the volume of the reaction vessel. End corrections are often necessary, since the reaction is not always immediately quenched the moment it leaves the reaction vessel at elevated temperatures. From the composition of the products, the rate of reaction can be determined, the procedure being repeated for a variety of different flow rates and with reaction vessels of different sizes.

The choice of a flow system to follow the reaction of carbon monoxide with ozone, using ozonized oxygen containing up to 10% ozone, is of some interest. The carbon monoxide and ozone both compete for oxygen atoms produced by the ozone de-

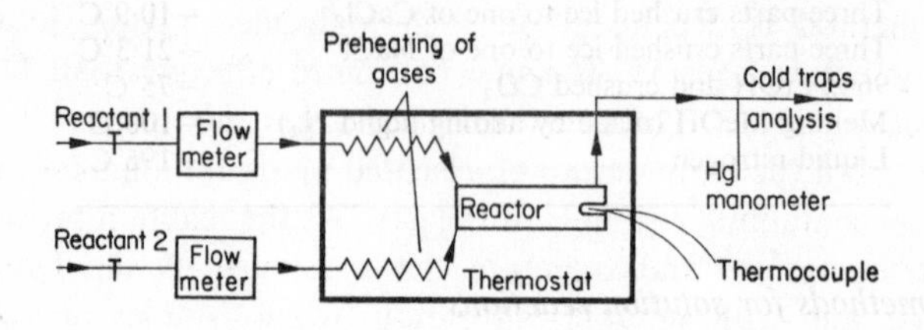

FIG. 8. Diagram of apparatus used for studying gas reactions by flow methods.

composition, but at temperatures up to 80°C, the amount of carbon dioxide produced is small. It becomes significant by working at higher temperatures in the range 160–290°C, but, under these conditions, the overall reaction is too fast to follow using static methods. Unreacted ozone is estimated by passing the gases through potassium iodide solutions and then titrating the liberated iodine. The concentration of carbon dioxide in the remaining carbon monoxide, oxygen, carbon dioxide gas mixture, was found by analysis of the 4·3 μ infrared absorption band.

Flow methods have also been used to detect low-yield products in reactions which are not necessarily fast. The separation and accumulation of these can usually be effected by passing the reacted gases through a series of cold traps in order of decreasing temperature. A range of suitable freezing mixtures is shown in

Table 1. When suitable amounts have been collected, the different products are identified and estimated quantitatively.

Flow methods use large quantities of reactants, and these are often hard to make and purify. Gas-phase chromatography can now be used to detect low concentration products (10^{-8} g), using a sample of reactant gases from a static reaction system.

TABLE 1

SOME FREEZING MIXTURES USED IN COLD TRAPS

Three parts crushed ice to one of $CaCl_2$	−10·9°C
Three parts crushed ice to one of NaCl	−21·3°C
96% EtOH and crushed CO_2	−75°C
Melting MeOH (made by adding liquid N_2)	−100°C
Liquid nitrogen	−196°C

Flow methods for solution reactions

In the continuous-flow method, the two reactant solutions are forced into the mixing chamber, and then through an observation tube with a velocity of several metres/second. The extent of the reaction can be determined by measuring the optical density, electrical conductivity or some other physical property at a fixed point along the observation tube, in which case a variety of flow rates are required. Alternatively, measurements can be made at various distances along the observation tube.

Continuous flow methods have been largely superseded by stopped-flow methods which use much less liquid (as little as 0·1–0·2 ml). The mixing is as before, but the flow is suddenly stopped, so that the solution comes to rest within a millisecond or two. The reaction in an element of solution at, say, 1 cm from the mixing chamber is then followed by making, in most cases, spectrophotometric observations at this point. The apparatus is illustrated in Fig. 9. The reactant solutions in syringes S_1 and S_2 are driven into a mixing chamber from which they pass into the observation tube O. The solution drives forward the light piston P until the latter reaches its seating Q, and, at this point, the flow of liquid is suddenly stopped. The moment it reaches Q, the marker N cuts off a light beam from L_1, and an electronic device, M_1, starts the reading of optical changes on the stationary solution in O (L_2 and M_2). Optical readings (with, for example, a 20 mm light path) can be recorded photographically by using oscillographic methods, or, for the slower reaction, with a pen recording chart. Reacted solutions are removed by way of W. The apparatus can be thermostatted over a 0–50°C range by

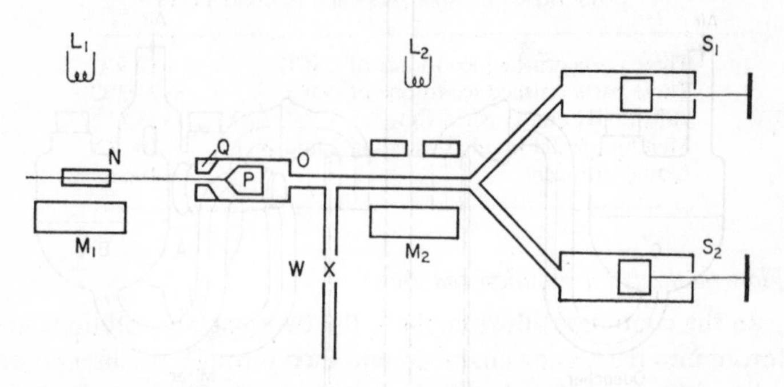

FIG. 9. The stopped flow method used for solution reactions. [Reproduced, with permission, from H. Q. GIBSON, *Discussions Faraday Soc.* **17**, 138 (1954).]

enclosing the syringes in steel blocks through which water is circulated, or by immersing the whole apparatus in a thermostat. Second-order rate constants in the 10^2–10^8 l mole^{-1} sec^{-1} range can be measured in this way. The method is as accurate as conventional techniques, the standard deviation being $\pm$1–2%.

Quenching techniques for fast reactions

Quenching techniques are particularly useful for studying isotopic exchange reactions just too rapid to be measured by conventional means, i.e. with half-lives of from 50 msec to 1–2 sec. The limiting factor is the rate at which the reaction solution can be quenched. Apparatus used to follow the exchange between MnO_4^{2-} and MnO_4^- using radioactive ^{54}Mn to label one of the reactants, is shown in Fig. 10. The procedure is briefly as follows. The reactants in A and B are allowed to reach thermal equilibrium, when an air pressure of 12 lb/in^2 is applied to A and B. The reactants are driven through 2 mm bore capillaries into the reaction

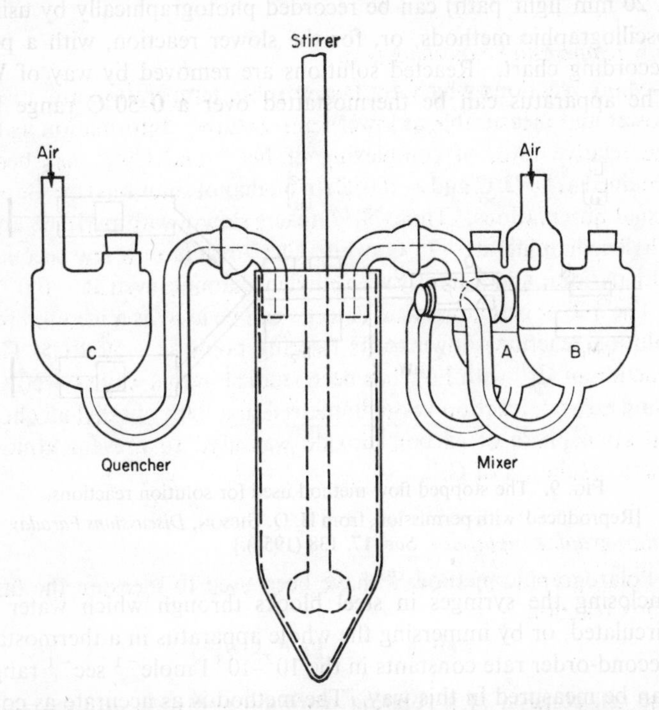

FIG. 10. Apparatus used to study the MnO_4^{2-}–MnO_4^- electron exchange using quenching techniques. [Reproduced with permission from J. C. SHEPPARD and A. C. WAHL, *J. Am. Chem. Soc.* **79**, 1021 (1957).]

vessel, and then, after a suitable time, a similar pressure is used to force the quenching solution in C into the reaction vessel. The quenching solution, in this case tetraphenylarsonium chloride, precipitates the permanganate ions. The latter are centrifuged off, and the activity of the ^{54}Mn determined. For each run, the above procedure is repeated 9–11 times, an electric timing device being used to vary the reaction time. Rate constants were in this particular case $\sim 10^3$ l mole^{-1} sec^{-1}, the standard deviation being $\pm 4\%$.

Low temperature experiments

Many reactions which are fast at room temperature are much slower and measurable at lower temperatures. Information as to the relative rates of complexing of Ni^{2+} and Cu^{2+} has been obtained at -75°C and -100°C (in methanol solutions) by simple visual observations. Thus, Ni^{2+} reacts slowly with pyridine and ethylenediamine at -75°C, while Cu^{2+} reacts in a few seconds with pyridine and instantly with ethylenediamine, even at -100°C.

The 5·27 N perchloric acid eutectic can be used as a medium for solution reactions down to its freezing point at $-59{\cdot}9$°C. The reaction of Cr^{II} with Fe^{III} has been studied in this way at -50°C, using a Pyrex reaction vessel immersed in a bath of ethyl alcohol. An atmosphere of carbon dioxide was used to prevent atmospheric oxidation of Cr^{II}.

Polarographic methods

Polarographic methods[6] have been used to measure the fast redox reaction

$$Fe^{II} + V^{V} \rightarrow Fe^{III} + V^{IV}.$$

The cell consists of a rotating platinum electrode in a reaction solution made up to 1 M in perchlorate, and this is connected by a salt-bridge to a standard calomel electrode. Only the Fe^{II} gives rise to a diffusion current when a potential of one volt is applied across the cell. As the Fe^{II} reacts, the diffusion current, which is proportional to the concentration, decreases and from a pen recorder trace second-order plots can be made. The rate constant at 0°C and in 1 N acid is $2{\cdot}4 \times 10^3$ l mole^{-1} sec^{-1}.

6. D. R. Rosseinsky, *Proc. Chem. Soc.* 16 (1963).

The much faster co-ordination reaction

$$Cd(CN)_3^- + CN^- \rightarrow Cd(CN)_4^{2-},$$

has been studied using a similar procedure. The rate constant is of the order 5×10^8 l mole^{-1} sec^{-1}.

Relaxation methods

Rate constants for reactions in solution are limited by the rate at which the reactants can diffuse together. For reactions between two uncharged reactants, the upper limit is in the region 10^9 l mole^{-1} sec^{-1}, while for a reaction between singly charged positive and negative ions, the limit is just greater than 10^{11} l mole^{-1} sec^{-1}. The problem is how to determine rate constants approaching these limiting values. All the methods so far described require a mixing of the reactants, and since this cannot be achieved in less than 10^{-3} sec, this generally excluded reactions with rate constants greater than 10^6 l mole^{-1} sec^{-1}.

The elegant relaxation techniques which have recently been pioneered by Eigen and colleagues[7] can be used for ionic reactions with half-lives ranging from 10^{-9} sec to over 1 sec. A rate constant of $k = 1{\cdot}5 \times 10^{11}$ l mole^{-1} sec^{-1} has, for example, been obtained for the reaction

$$H^+ + OH^- \rightarrow H_2O.$$

Features of such methods are as follows. *The reaction system must first of all be one in which there is an observable equilibrium.* Since the equilibrium is dependent on pressure, electric field intensity and temperature, it will be disturbed if one of these quantities is suddenly changed. The rapid approach of the new equilibrium can be followed using a cathode-ray oscillograph to register spectroscopic or conductance changes, the trace obtained on the oscillograph being recorded photographically.

In the temperature-jump method, a temperature jump of the

7. M. EIGEN, *Advances in the Chemistry of the Coordination Compounds* p. 371 (Edited by KIRSCHNER), MacMillan, New York, 1961.

order of 2–10°C can be effected in about 10^{-6} sec by the discharge from a high voltage condenser at a potential difference of the order of 100 kV. From resultant concentration–time plots, relaxation times and corresponding rate constants can be obtained. The kinetics of the electron-transfer reaction

$$Fe(DMP)_3^{2+} + IrCl_6^{2-} \underset{k_{-1}}{\overset{k_1}{\rightleftarrows}} Fe(DMP)_3^{3+} + IrCl_6^{3-},$$

where DMP = 4,7-dimethyl-1,10-phenanthroline, has been studied in this way.[8] At 10°C $k_1 = 1{\cdot}1 \times 10^9$ l mole^{-1} sec^{-1} and $k_{-1} = 1{\cdot}0 \times 10^9$ l mole^{-1} sec^{-1}.

Alternatively, the physical quantity can be rapidly and periodically varied, and this method is generally to be preferred, since the application of an alternating high-density electric field, or the incidence of ultrasonic waves to bring about periodic fluctuations in temperature and pressure, is often much more convenient. The reaction will lag behind the rapidly changing physical parameter, and a net observable displacement of the equilibrium will result. Using a mathematical treatment, the observed shifts can be related to the chemical rate constants.

The ultrasonic relaxation method has been used to study the rate of complexing of SO_4^{2-} with 2+ metal ions.[9] In each case, the results indicate a stepwise formation of the inner-sphere complex, the first stage being the diffusion controlled formation of an ion-pair or outer-sphere complex,

$$M(H_2O)_6^{2+} + SO_4^{2-} \rightleftharpoons M(H_2O)_6^{2+}, SO_4^{2-}.$$

In the final stage, there is rearrangement of the ion-pair to give the inner-sphere complex,

$$M(H_2O)_6^{2+}, SO_4^{2-} \underset{k_{-1}}{\overset{k_1}{\rightleftarrows}} M(H_2O)_5\,SO_4 + H_2O.$$

---

8. J. Halpern, R. J. Legare and R. Lumry, *J. Am. Chem. Soc.* **85**, 680 (1963).
9. M. Eigen, *Z. Elektrochem.* **64**, 115 (1960).

Rate constants for k_1 and k_{-1} were evaluated and are shown in Table 2.

TABLE 2

RATE CONSTANTS FOR THE FORMATION AND SUBSTITUTION OF SULPHATE COMPLEXES AT 25°C

M	$k_1(\text{sec}^{-1})$	$k_{-1}(\text{sec}^{-1})$
Be^{2+}	1×10^2	$1{\cdot}3 \times 10^3$
Mg^{2+}	1×10^5	8×10^5
Mn^{2+}	4×10^6	2×10^7
Co^{2+}	2×10^5	$2{\cdot}5 \times 10^6$
Ni^{2+}	1×10^{-4}	1×10^5
Cu^{2+}	(10^4)	1×10^6

Flash photolysis

In flash photolysis, the reactants (in the gas phase or in solution) are subject to an intense burst of light energy (duration $\sim 10^{-5}$ sec) such that energies of up to 10^5 joules can be absorbed in a few microseconds (10^{-6}–10^{-4} sec). With suitable reactants, the absorption initiates a decomposition, and free-radicals are formed in relatively high concentrations. Subsequent reactions of the free-radicals can be studied using absorption spectroscopy.

In a typical procedure, the reaction vessel is a cylindrical quartz vessel some 50 cm long and 2 cm in diameter, the photolysis flash tube being placed alongside to achieve uniform radiation. To obtain a complete spectrum of the radicals produced, a second spectroscopic flash of lower energy can be used, the whole of the spectrum being recorded on a photographic plate. The timing of the two flashes is electronically controlled so that from a series of experiments in which the time is varied, the rate of disappearance of the radicals can be studied.

For kinetic work, photocells allow a much more accurate measurement of the light intensity to be made. If the reaction vessel is subject to a continuous spectroscopic source, such as might be obtained from a tungsten lamp, and the light is spectrographed after passing through the reaction vessel (Fig. 11), slits can then be set in front of photomultipliers of chosen wavelengths. The current obtained at a particular wavelength can be amplified and fed to a cathode-ray oscillograph, the time-base of which is triggered by the initial flash. The oscillograph gives a trace showing the initial sharp formation of the radical followed

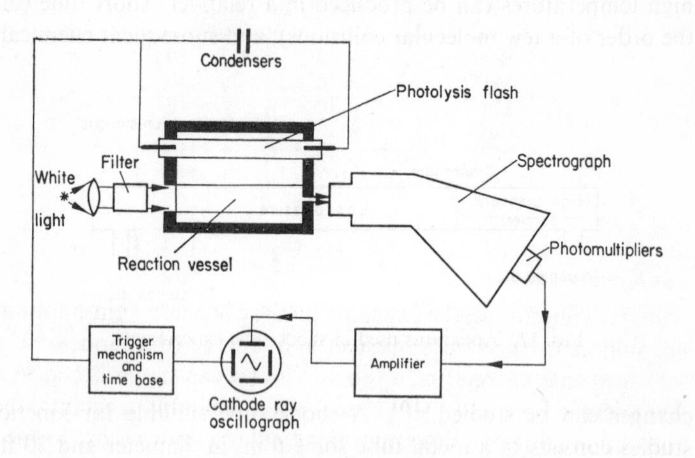

FIG. 11. Kinetic studies using flash photolysis.

by its subsequent decay. The whole of the trace can be photographed, the rate of disappearance of the radical being of particular interest.

The iodine-atom recombination reaction has been extensively studied in this way. In such experiments, the temperature of the reaction vessel is controlled by an outer jacket filled with silicone oil, a heating element allowing the temperature to be varied between 20 and 250°C. By surrounding the reaction vessel with a saturated solution of copper sulphate, the light of longer wavelengths from the initial flash is absorbed, thus reducing any unwanted heating effect. Some 20% dissociation of iodine can result by flashing up to 0·15 mm of iodine in 100 mm of inert gas. The recombination is followed by measuring the reformation of molecular iodine at 520 mμ. Reactions are complete within 10^{-2} sec. Third-order rate constants as high as 10^{11} l^2 $mole^{-2}$ sec^{-1} have been measured.

Shock tubes

Shock tubes are particularly useful in gas-phase work, since high temperatures can be produced in a relatively short time (of the order of a few molecular collisions), and subsequent chemical

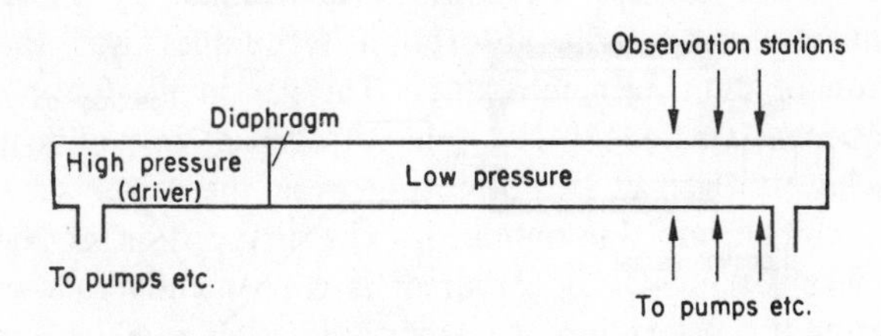

FIG. 12. Apparatus used in shock-tube experiments.

changes can be studied.[10] A shock tube suitable for kinetic studies consists of a metal tube some 6 in. in diameter and 20 ft long (Fig. 12). This is divided into a high-pressure section, containing driver gas, and a low-pressure section containing the gas or gases to be reacted. The two are separated by a thin diaphragm of metal foil or cellophane. When the latter is burst by a suitably controlled needle, the expansion of the high-pressure gas produces a shock wave which passes through the low-pressure gases. These are adiabatically compressed and are carried to the far end of the low-pressure section where subsequent changes can be followed (observation stations).

With an inert gas, the shock wave produces a heating effect and then thermal ionization. If the low-pressure section contains a diatomic gas, other things can happen. Starting with a weak

10. H. O. PRITCHARD, *Quart. Rev.* **14,** 48 (1960).

shock wave, travelling only a little faster than sound, changes in rotational and vibrational energies can be observed. The gas is suddenly heated to a new temperature, but it takes many collisions before the molecular rotations and vibrations come to thermal equilibrium with the translational motion. With stronger shock waves, temperatures are reached at which the diatomic molecule begins to dissociate into radicals, and since this is a relatively slow process, rates can often be measured. At higher temperatures still, ionization can occur, and rates of ionization and recombination can be measured.

One of the earliest reactions to be studied by shock-tube experiments using light absorption techniques was the dissociation of dinitrogen tetroxide. The gas in the low-pressure compartment was ~1% N_2O_4 in nitrogen at one atmosphere, and using nitrogen at two atmospheres as the driver gas, a 25° rise in temperature was obtained. The dissociation of N_2O_4 to NO_2, where the NO_2 is coloured, is conveniently followed by measuring the absorption of mercury radiation at 400 mμ.

The rate of dissociation of halogen molecules has been studied by shock-wave techniques. In certain cases, it is possible to initiate the formation of radicals from one of the molecular species, and to study the subsequent reaction of these with a second molecular species. The reactions

$$Br + H_2 \rightarrow HBr + H$$

and

$$H + O_2 \rightarrow OH + O,$$

have been studied in this way, since Br_2 (in a mixture of Br_2 and H_2) and H_2 (in a mixture of H_2 and O_2) have the lower dissociation energies.

Ion–molecule reactions

The occurrence of ions in the gas phase is rare, and only under rather extreme conditions as, for example, in an electric discharge tube, or with the incidence of α-, β- and γ-rays, are they readily produced. They are also formed in the ionization chamber of a mass spectrometer, and, under certain conditions, the secondary reactions of such ions can be followed.

The three stages in mass spectrometry are (*a*) the production of ions, (*b*) the separation of ions according to their mass/charge values and (*c*) the measurement of ion concentrations corresponding to their different mass/charge peaks. Ions are produced in the ionization chamber by the impact of electrons accelerated by direct voltages of up to 100 V. The positive ions produced are

TABLE 3

SECOND ORDER RATE CONSTANTS FOR SOME ION–MOLECULE REACTIONS

Reaction	Rate constant ($l\ mole^{-1}\ sec^{-1}$)
$He^+ + H_2 \rightarrow HeH^+ + H$	$0{\cdot}37 \times 10^{12}$
$Ar^+ + H_2 \rightarrow ArH^+ + H$	$1{\cdot}01 \times 10^{12}$
$H_2^+ + H_2 \rightarrow H_3^+ + H$	$1{\cdot}70 \times 10^{12}$
$H_2^+ + O_2 \rightarrow HO_2^+ + H$	$4{\cdot}60 \times 10^{12}$
$HCl^+ + HCl \rightarrow H_2Cl + Cl$	$0{\cdot}27 \times 10^{12}$

subsequently expelled into an accelerating field (1000 V) and the various peaks measured in the mass analyser (they are converted into small peak currents). If, however, the pressure of gases in the ionization chamber is not sufficiently low, the ions first formed will react to give secondary ions. Such ions with their corresponding mass peaks are generally unwanted, and, in modern mass spectrometry, fewer peaks and simpler patterns are generally obtained by working at pressures $<10^{-4}$ mm Hg.

However, by working at pressures of around 10^{-3} mm, the rate of formation of secondary ions can be studied. The ratio of mass spectral currents for the primary and secondary ions at a series of pressures in this region are measured. A review of the subject and details of the method used to evaluate specific rate constants is given in Ref. 11. Examples of reactions studied are shown in Table 3.

11. F. W. LAMPE, J. L. FRANKLIN and F. H. FIELD, *Progress in Reaction Kinetics* (Edited by PORTER), Pergamon Press, London, 1961.

ESR and NMR spectroscopy

Electron-spin resonance is a phenomenon observed when magnetic fields are applied to metal ions or free-radicals possessing unpaired electrons. Information regarding the spacial distribution of such electrons can be obtained if they move in the vicinity of an atom or atoms with nuclear spins, since the resonance spectrum then has a fine structure which is characteristic of that system. In electron-transfer reactions, the movement of an odd electron from one reactant to another results in a line broadening from which rate constants can be estimated. The method can be used to measure second-order rate constants in 10^5–10^{10} l mole^{-1} sec^{-1} range, but, as yet, has not been extensively used in the field of inorganic reactions. The upper limit of $k < 4 \times 10^8$ l mole^{-1} sec^{-1} for the electron-transfer reaction of $W(CN)_8^{4-}$ with $W(CN)_8^{3-}$ is consistent with previous estimates of $k > 4 \times 10^4$ l mole^{-1} sec^{-1} using radioactive tracers.[12]

A still wider range of rate constants can be measured using nuclear magnetic resonance techniques, which are concerned with the response of nuclear spins in different electron environments to applied magnetic fields. The electron-transfer reactions between Cu^{I} and Cu^{II}, and V^{IV} and V^{V}, have for example been studied using the ^{63}Cu and ^{51}V isotopes which have nuclear spins. The rate of exchange between Cu^{I} and Cu^{II} was determined from the extent of broadening of the ^{63}Cu NMR spectrum as Cu^{2+}(10^{-4}–10^{-3} M) was added to a solution of 0·1 M CuCl in 12 M HCl.

12. S. I. Weissman and C. S. Garner, *J. Am. Chem. Soc.* **78,** 1072 (1956).
13. H. M. McConnell and S. B. Berger, *J. Chem. Phys.* **27,** 230 (1953).

CHAPTER 3

TREATMENT OF DATA

THE elementary reactions which together make up a reaction mechanism are almost invariably bimolecular processes. Once the stoicheiometry and products of an overall reaction are known, it is generally possible to propose more than one possible mechanism, depending, of course, on the complexity of the system. In choosing between these existing kinetic information for related systems, the effect which products, and in some cases substrates, have on the rate, and the identity of reaction intermediates are often a great help. For a mechanism to be finally accepted, however, the concentration–time data from kinetic experiments must fit the rate equation and, if integration is possible, the integrated rate equation for that mechanism. Even then, although a mechanism may be accepted because it gives a satisfactory account of a given set of experimental data, with further experiments it may have to be amended.

Simple rate equations

Many reactions with overall equations

$$A + B \rightarrow C + D$$

obey second-order kinetics,

$$\frac{-d[A]}{dt} = \frac{-d[B]}{dt} = \frac{d[C]}{dt} = \frac{d[D]}{dt} = k[A][B],$$

where [A] and [B] are reactant concentrations at a time t. The rate constant k can be evaluated by measuring gradients, $-d[A]/dt$, to a concentration–time curve such as that shown in

Fig. 13. A better method, however, i.e. one showing greater accuracy, is to use the integrated rate equation, which in this case is

$$\log_{10}\frac{[A]}{[B]} = \frac{([A]_0 - [B]_0)}{2{\cdot}303}kt + \log_{10}\frac{[A]_0}{[B]_0}$$

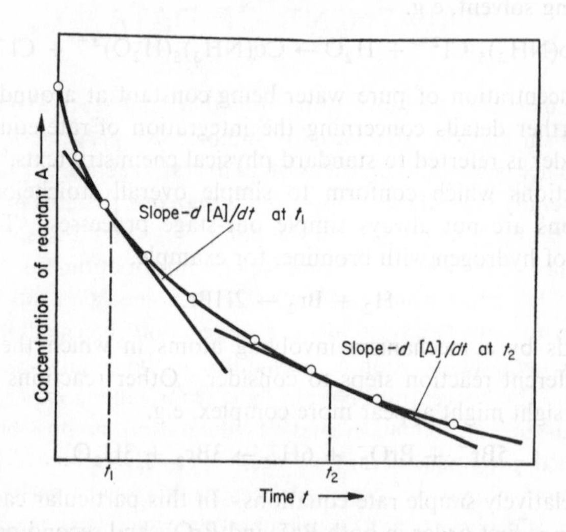

FIG. 13. The rate of a reaction at a given time can be obtained by drawing a tangent to the concentration–time curve.

where $[A]_0$ and $[B]_0$ are initial reactant concentrations. Thus by plotting $\log_{10}([A]/[B])$ against t, the rate constant can be obtained from the gradient $([A] - [B])\, k/2{\cdot}303$.

When one of the reactants is present in large excess, at least a ten-fold excess, the concentration of that reactant can be assumed to remain constant throughout a run, and the reaction is said to be pseudo first-order. If in the above case B is in large excess, then

$$\frac{-d[A]}{dt} = k'[A], \text{ where } k' = k[B].$$

On integration

$$\log_{10}[A] = \frac{-k'}{2 \cdot 303} t + \log_{10}[A]_0$$

so that k' can be obtained from linear plots of $\log_{10}[A]$ against t. An obvious example of a pseudo first-order reaction is one involving solvent, e.g.

$$Co(NH_3)_5 Cl^{2+} + H_2O \rightarrow Co(NH_3)_5(H_2O)^{3+} + Cl^-,$$

the concentration of pure water being constant at around 55 M. For further details concerning the integration of rate equations the reader is referred to standard physical chemistry texts.[1]

Reactions which conform to simple overall stoicheiometric equations are not always simple one-stage processes. The reaction of hydrogen with bromine, for example,

$$H_2 + Br_2 \rightarrow 2HBr,$$

proceeds by a mechanism involving atoms in which there are five different reaction steps to consider. Other reactions which at first sight might appear more complex, e.g.

$$5Br^- + BrO_3^- + 6H^+ \rightarrow 3Br_2 + 3H_2O,$$

have relatively simple rate equations. In this particular case, the reaction is first-order in both Br^- and BrO_3^- and second order in $[H^+]$, i.e.

$$\frac{-d[Br^-]}{dt} = k[Br^-][BrO_3^-][H^+]^2,$$

which is consistent with a mechanism

$$BrO_3^- + 2H^+ \overset{\text{fast}}{\rightleftharpoons} H_2BrO_3^+ \quad (1)$$

$$Br^- + H_2BrO_3^+ \rightarrow Br\text{—}BrO_2 + H_2O \quad (2)$$

$$Br\text{—}BrO_2 + 4H^+ + 4Br^- \overset{\text{fast}}{\rightarrow} \text{products}, \quad (3)$$

in which reaction (2) is rate determining.

1. Cf. W. J. MOORE, *Physical Chemistry*, 4th ed., pp. 260–70, Longmans, London, 1963.

Although a few reactions show good third-order behaviour, it is unlikely that these occur in one termolecular step, since such reactions are improbable. The reaction

$$2NO + O_2 \rightarrow 2NO_2,$$

for example, obeys a rate equation

$$\frac{-d[NO]}{dt} = k[NO]^2[O_2],$$

under all conditions so far studied, but is believed to proceed with the intermediate formation of either $(NO)_2$ or NO_3 (or both). In such reactions (and many others besides), care is required in defining the experimental rate constant, since k, in the above case, is twice that defined by the rate equation

$$\frac{-d[O_2]}{dt} = k[NO]^2[O_2].$$

Fractional orders are generally characteristic of heterogeneous gas reactions at a solid surface, the reactant or one of the reactants being only moderately absorbed. Thus, in the decomposition of stibine (SbH_3) at an antimony surface, the order of the reaction is 0·6. In the reaction between carbon monoxide and oxygen at a quartz surface, on the other hand,

$$2CO + O_2 \rightarrow 2CO_2,$$

carbon monoxide being strongly absorbed and oxygen only weakly absorbed, there is an inverse dependence on [CO], and the rate equation is of the form

$$\frac{-d[CO]}{dt} = \frac{k[O_2]}{[CO]},$$

where $[O_2]$ and [CO] are gas-phase concentrations.

For a reaction of unknown mechanism, a first approach might be to assume that the rate equation takes the simple form

$$\frac{-d[A]}{dt} = k[A]^m[B]^n,$$

i.e. it is dependent on reactant concentrations only. If then, B, say, is taken in large excess or alternatively if A is the only reactant,

$$\frac{-d[\mathrm{A}]}{dt} = k'[\mathrm{A}]^m.$$

The latter may be rewritten

$$-\log_{10}(d[\mathrm{A}]/dt) = m \log_{10}[\mathrm{A}] + \log_{10}k'$$

and since the rate, $d[\mathrm{A}]/dt$, can be obtained by drawing gradients to concentration–time curves (Fig. 13), m can be determined by measuring the gradient of $\log_{10}(d[\mathrm{A}]/dt)$ against $\log_{10}[\mathrm{A}]$ plots. To determine n, a similar procedure is used with [A] in excess.

Equilibrium reactions

A well-known equilibrium reaction is that involving hydrogen iodide, hydrogen and iodine,

$$2\mathrm{HI} \rightleftharpoons \mathrm{H_2} + \mathrm{I_2}.$$

At temperatures up to 300°C, the contribution made by atom reactions is negligible, and the forward and back reactions are essentially single-stage bimolecular processes as shown.(2) In the decomposition of hydrogen iodide to an equilibrium mixture, the rate may be expressed as

$$\frac{-d[\mathrm{HI}]}{dt} = k_1[\mathrm{HI}]^2 - k_{-1}[\mathrm{H_2}][\mathrm{I_2}],$$

or

$$\frac{dx}{dt} = k_1(a - x)^2 - k_{-1}(x/2)^2,$$

where a is the initial concentration of hydrogen iodide and x the amount which has reacted after a time t. At equilibrium

$$k_1(a - x_e)^2 = k_{-1}(x_e/2)^2,$$

2. M. Bodenstein, *Z. Physik. Chem.* **29**, 295 (Leipzig, 1899).

and substituting for k_{-1},

$$\frac{-dx}{dt} = k_1(a-x)^2 - \frac{k_1(a-x_e)^2\, x^2}{x_e^2}.$$

On integration

$$\log_{10} \frac{x(a-2x_e)+ax_e}{a(x_e-x)} = \frac{2k_1at(a-x_e)}{2{\cdot}303x_e},$$

and k can be evaluated in the usual way. The reverse reaction can be studied in a similar fashion, but since it is not easy to seal up equivalent amounts of hydrogen and iodine, the rate equation generally takes the form

$$\frac{dx}{dt} = k_{-1}(a-x)(b-x) - k_1(x/2)^2,$$

where $x/2$ is the amount of HI formed at a time t. The corresponding integrated equation is, in this case, more complex.

The effect of back reactions: The stationary-state approximation

For systems having more than one step, back reactions often play a significant role in determining the progress of the overall reaction. Thus a back reaction will be effective if its rate is of the same order of magnitude or greater than that of a subsequent forward reaction, where the two have a reactant in common. Often the latter is a free-radical intermediate, but this need not necessarily be so, and, in certain instances, the intermediate is a perfectly stable chemical. Because of the competition for, and rapid reaction of such intermediates, it often happens that the concentration of the intermediate remains small. If this is so, it often follows that the rate of formation and disappearance of the intermediate in moles per unit time can never be other than small, compared to the rate of disappearance of the principal reactants and the rate of formation of products. Thus, for an intermediate I, we can say

$$\frac{d[I]}{dt} = 0,$$

or, in other words, we can assume that the concentration of the intermediate remains constant. If the time required for the initial formation of the intermediate is slow, an induction period is observed and the steady-state approximation is only valid when this period is complete.

The stationary-state approximation is invaluable in deriving rate equations for multi-stage reactions. Its net effect is generally to introduce reactant and product concentration terms into the denominator of rate equations. Consider, for example, the reaction between hydrogen and bromine,[3]

$$H_2 + Br_2 \rightarrow 2HBr.$$

The mechanism is believed to be

$$Br_2 + M \underset{k_{-1}}{\overset{k_1}{\rightleftharpoons}} 2Br + M$$

$$Br + H_2 \underset{k_{-2}}{\overset{k_2}{\rightleftharpoons}} HBr + H$$

$$H + Br_2 \overset{k_3}{\rightarrow} HBr + Br,$$

where M is a third-body which provides some of the energy which is necessary to break the Br–Br bond. The rate equation can be derived by making two stationary-state approximations,

$$\frac{d[Br]}{dt} = 0 \text{ and } \frac{d[H]}{dt} = 0,$$

so that

$$\frac{d[Br]}{dt} = k_1[Br_2][M] - k_{-1}[Br]^2[M] - k_2[H_2][Br] + k_{-2}[HBr][H] + k_3[H][Br_2] = 0$$

3. Cf. M. Polanyi, *Z. Elektrochem.* **26**, 49 (1920).

and

$$\frac{d[\mathrm{H}]}{dt} = k_2[\mathrm{H}_2][\mathrm{Br}] - k_{-2}[\mathrm{HBr}][\mathrm{H}] - k_3[\mathrm{H}][\mathrm{Br}_2] = 0.$$

By solving for [H] and [Br], and substituting in the overall rate expression

$$\frac{d[\mathrm{HBr}]}{dt} = k_2[\mathrm{H}_2][\mathrm{Br}] - k_{-2}[\mathrm{HBr}][\mathrm{H}] + k_3[\mathrm{H}][\mathrm{Br}_2],$$

a rate equation

$$\frac{d[\mathrm{HBr}]}{dt} = \frac{2(k_1/k_{-1})^{1/2}\, k_2 k_3 [\mathrm{H}_2][\mathrm{Br}_2]^{3/2}}{k_3[\mathrm{Br}_2] + k_{-2}[\mathrm{HBr}]}$$

is obtained.

Hence

$$\frac{[\mathrm{H}_2][\mathrm{Br}_2]^{1/2}}{d[\mathrm{HBr}]/dt} = \frac{k_{-2}[\mathrm{HBr}]}{2k_2k_3(k_1/k_{-1})^{1/2}\,[\mathrm{Br}_2]} + \frac{1}{2k_2(k_1/k_{-1})^{1/2}}$$

and straight line graphs are obtained when the left-hand side of this equation is plotted against $[\mathrm{HBr}]/[\mathrm{Br}_2]$.

Other reaction systems will now be considered in somewhat greater detail.

The reaction between iron(II) and thallium(III)

The reaction of iron(II) with thallium(III),

$$2\mathrm{Fe}^{\mathrm{II}} + \mathrm{Tl}^{\mathrm{III}} \rightarrow 2\mathrm{Fe}^{\mathrm{III}} + \mathrm{Tl}^{\mathrm{I}},$$

is said to be non-complementary, since the stable oxidation states of iron differ by one electron, and those of thallium (outer electron configuration $6s^2 6p^1$) by two. In considering possible mechanisms,[4] termolecular reactions are first of all unlikely on grounds of probability. The two mechanisms

$$\mathrm{Fe}^{\mathrm{II}} + \mathrm{Tl}^{\mathrm{III}} \xrightarrow{\text{slow}} \mathrm{Fe}^{\mathrm{III}} + \mathrm{Tl}^{\mathrm{II}}$$

$$\mathrm{Fe}^{\mathrm{II}} + \mathrm{Tl}^{\mathrm{II}} \xrightarrow{\text{fast}} \mathrm{Fe}^{\mathrm{III}} + \mathrm{Tl}^{\mathrm{I}},$$

4. K. G. Ashurst and W. C. E. Higginson, *J. Chem. Soc.* 3044 (1953).

and

$$Fe^{II} + Tl^{III} \xrightarrow{\text{slow}} Fe^{IV} + Tl^{I}$$

$$Fe^{II} + Fe^{IV} \xrightarrow{\text{fast}} 2Fe^{III},$$

involving a slow and fast step only, are, moreover, clearly inadequate, since the simple rate equation

$$\frac{-d[Fe^{II}]}{dt} = k[Fe^{II}][Tl^{III}],$$

is not obeyed over a sufficiently wide range, i.e. to at least 90% completion. Second-order plots of $\log_{10}[Fe^{II}]/[Tl^{III}]$ against time are, in fact, linear to around 60% completion, after which they become curved. The direction of the curvature suggests that a back reaction is becoming effective, and the above reaction sequences can be amended accordingly to read

$$Fe^{II} + Tl^{III} \rightleftharpoons Fe^{III} + Tl^{II}$$

$$Fe^{II} + Tl^{II} \rightarrow Fe^{III} + Tl^{I},$$

and

$$Fe^{II} + Tl^{III} \rightleftharpoons Fe^{IV} + Tl^{I}$$

$$Fe^{IV} + Fe^{II} \rightarrow 2Fe^{III}.$$

When this reaction was first studied, there was no independent evidence for the existence of either of the aqueous ions Fe^{IV} or Tl^{II}, and there was no reason for favouring one or other of these mechanisms. By inspection, however, Fe^{III} and not Tl^{I} would be expected to slow down the first, and Tl^{I} but not Fe^{III} to slow down the second of these mechanisms. For solutions around 10^{-3} M in reactants, the addition of a thirty-fold excess of Fe^{III} produces a marked retardation (second-order plots become curved at a much earlier stage), while similar amounts of Tl^{I} have no effect. Thus the reaction would appear to be largely, if not exclusively, by the mechanism

$$Fe^{II} + Tl^{III} \underset{k_{-1}}{\overset{k_1}{\rightleftharpoons}} Fe^{III} + Tl^{II}$$

$$Fe^{II} + Tl^{II} \xrightarrow{k_2} Fe^{III} + Tl^{I}.$$

Assuming stationary-state kinetics for Tl^{II},

$$\frac{d[Tl^{II}]}{dt} = k_1[Fe^{II}][Tl^{III}] - k_{-1}[Fe^{III}][Tl^{II}] - k_2[Fe^{II}][Tl^{II}] = 0$$

and

$$[Tl^{II}] = \frac{k_1[Fe^{II}][Tl^{III}]}{k_{-1}[Fe^{III}] + k_2[Fe^{II}]}.$$

But the rate of formation of Tl^{I} is

$$\frac{d[Tl^{I}]}{dt} = k_2[Fe^{II}][Tl^{II}],$$

and since

$$\frac{-2d[Tl^{III}]}{dt} = \frac{-d[Fe^{II}]}{dt} = \frac{2d[Tl^{I}]}{dt} = \frac{d[Fe^{III}]}{dt},$$

the overall ratio is given by

$$\frac{-d[Fe^{II}]}{dt} = 2k_2[Fe^{II}][Tl^{II}].$$

Hence, substituting for $[Tl^{II}]$

$$\frac{-d[Fe^{II}]}{dt} = \frac{2k_1k_2[Fe^{II}]^2[Tl^{III}]}{k_{-1}[Fe^{III}] + k_2[Fe^{II}]}.$$

Notice that the two reactants competing for Tl^{II} are represented on the denominator of the rate equation.

In order to evaluate rate constants, the rate $-d[Fe^{II}]/dt$ may be obtained by drawing tangents to a concentration–time curve, or alternatively (and preferably), the above rate equation may be integrated. On integration a rather unwieldy expression

$$\frac{k_{-1}(a+c)}{k_2 - k_{-1}\left(1+\dfrac{a+c}{b-a}\right)}\frac{1}{[\mathrm{Fe^{II}}]} - \log_e \frac{[\mathrm{Fe^{II}}]}{2[\mathrm{Tl^{III}}]} =$$

$$\frac{k_2(b-a)}{k_2 - k_{-1}\left(1+\dfrac{a+c}{b-a}\right)} k_1 t + M.$$

is obtained, where a, b and c are the initial concentrations of Fe^{II}, $Tl^{III} \times 2$ and Fe^{III} respectively, and M is a constant. The procedure then is to substitute possible k_{-1}/k_2 values into this equation until a plot of the left-hand side against t gives a straight line (a procedure which is remarkably sensitive and by no means as tedious as it may sound). It is also possible to obtain k_{-1}/k_2 values using an alternative graphical method, but this yields less accurate values than the above. Whichever method is used, it is only possible to obtain k_1 and the ratio k_{-1}/k_2.

The reaction between the mercury(I) dimer and thallium(III)

In this reaction Hg^{II} but not Tl^{I} produces a retardation[5] which is in keeping with a mechanism

$$(\mathrm{Hg^I})_2 \underset{k_{-1}}{\overset{k_1}{\rightleftharpoons}} \mathrm{Hg^0} + \mathrm{Hg^{II}} \tag{1}$$

$$\mathrm{Hg^0} + \mathrm{Tl^{III}} \xrightarrow{k_2} \mathrm{Hg^{II}} + \mathrm{Tl^{I}}. \tag{2}$$

Here, Hg^0 implies the mercury atom, evidence for the existence of which in aqueous solutions ($\sim 10^{-7}$ M at 20°C) comes from the direct measurement of the solubility of metallic mercury in water,

5. A. M. Armstrong, J. Halpern and W. C. E. Higginson, *J. Phys. Chem.* 60, 1661 (1956).

and from spectroscopic observation of the 254 mμ mercury absorption. Assuming stationary-state kinetics for Hg^0,

$$\frac{d[Hg^0]}{dt} = k_1([Hg^I]_2) - k_{-1}[Hg^{II}][Hg^0] - k_2[Hg^0][Tl^{III}] = 0,$$

so that

$$[Hg^0] = \frac{k_1[(Hg^I)_2]}{k_{-1}[Hg^{II}] + k_2[Tl^{III}]}$$

and from eqn.(2), the rate can be expressed

$$\frac{-d[Tl^{III}]}{dt} = \frac{k_1 k_2 [Tl^{III}]([Hg^I]_2)}{k_{-1}[Hg^{II}] + k_2[Tl^{III}]}.$$

But the experimental results fit an empirical rate equation of the form

$$\frac{-d[Tl^{III}]}{dt} = \frac{a[Tl^{III}]([Hg^I]_2)}{[Hg^{II}]}$$

where a is a constant; and only by assuming that $k_{-1}[Hg^{II}] >> k_2[Tl^{III}]$ are the two in agreement. Alternatively, we may begin by making this same assumption and treating the equilibrium step

$$(Hg^I)_2 \underset{k_{-1}}{\overset{k_1}{\rightleftharpoons}} Hg^{II} + Hg^0$$

as a fast pre-equilibrium. Thus,

$$[Hg^0] = \frac{k_1([Hg^I]_2)}{k_{-1}[Hg^{II}]},$$

and from equation (2)

$$\frac{-d[Tl^{III}]}{dt} = \frac{k_1 k_2 [Tl^{III}]([Hg^I]_2)}{k_{-1}[Hg^{II}]}.$$

The decomposition of nitrogen dioxide

In some cases, the reaction products can be formed by two quite distinct routes, and the rate is the arithmetic sum of two concentration terms. The thermal decomposition of NO_2 is a good example.

Until recently, the reaction was interpreted in terms of a simple bimolecular step,

$$2NO_2 \rightarrow 2NO + O_2.$$

A more careful investigation has recently shown that rates are faster over the earlier stages of the reaction, and at 400°C approximately twice those expected from the simple bimolecular step. After ~10% decomposition, the reaction then conforms to the simple rate equation

$$\frac{d[NO_2]}{dt} = k_{obs}[NO_2]^2,$$

i.e. second-order plots become linear. The faster reaction over the initial stages can be accounted for by considering a mechanism

$$2NO_2 \xrightarrow{k_1} 2NO + O_2,$$

$$2NO_2 \underset{k_{-2}}{\overset{k_2}{\rightleftharpoons}} NO_3 + NO,$$

$$NO_3 + NO_2 \xrightarrow{k_3} NO + O_2 + NO_2.$$

Making the stationary-state approximation for $[NO_3]$

$$\frac{d[NO_3]}{dt} = k_2[NO_2]^2 - k_{-2}[NO_3][NO] - k_3[NO_3][NO_2] = 0,$$

and from this

$$[NO_3] = \frac{k_2[NO_2]^2}{k_{-2}[NO] + k_3[NO_2]}.$$

But

$$\frac{d[O_2]}{dt} = k_1[NO_2]^2 + k_3[NO_2][NO_3],$$

and since

$$\frac{2d[O_2]}{dt} = \frac{d[NO]}{dt} = \frac{-d[NO_2]}{dt},$$

$$\frac{-d[NO_2]}{2dt} = k_1[NO_2]^2 + k_3[NO_2][NO_3].$$

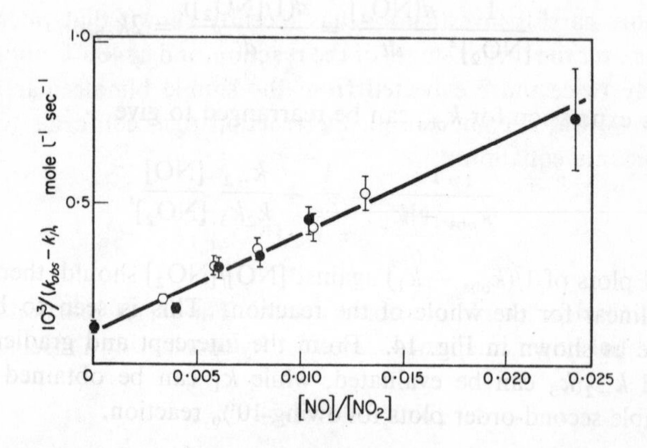

FIG. 14. Plot of $1/(k_{obs} - k_1)$ against $[NO]/[NO_2]$ for the decomposition of nitrogen dioxide at 411°C. Open circles are for points during a run; shaded circles with NO added. [Reproduced, with permission, from P. G. ASHMORE and M. G. BURNETT, *Trans. Faraday Soc.* **58**, 253 (1962).]

Substituting for $[NO_3]$,

$$\frac{-d[NO_2]}{2dt} = k_1[NO_2]^2 + \frac{k_2k_3[NO_2]^3}{k_{-2}[NO] + k_3[NO_2]}.$$

If then an experimental rate equation

$$-\frac{d[NO_2]}{2dt} = k_{obs}[NO_2]^2$$

is used, k_{obs} is given by

$$k_{obs} = k_1 + \frac{k_2k_3[NO_2]}{k_{-2}[NO] + k_3[NO_2]}.$$

When significant amounts of nitric oxide are formed, i.e. after 10% reaction, $k_{-2}[NO] \gg k_3[NO_2]$, and the second term in this expression becomes negligible.

In the initial stages k_{obs} can be obtained by drawing gradients to a plot of $1/[NO_2]$ against time, where

$$\frac{1}{[NO_2]^2}\frac{d[NO_2]}{dt} = \frac{d(1/[NO_2])}{dt} = 2k_{obs}.$$

The expression for k_{obs} can be rearranged to give

$$\frac{1}{k_{obs} - k_1} = \frac{1}{k_2} + \frac{k_{-2}}{k_2 k_3}\frac{[NO]}{[NO_2]},$$

and plots of $1/(k_{obs} - k_1)$ against $[NO]/[NO_2]$ should, therefore, be linear for the whole of the reaction. This is seen to be the case as shown in Fig. 14. From the intercept and gradient, k_2 and k_{-2}/k_3 can be evaluated, while k_1 can be obtained from simple second-order plots following 10% reaction.

Pre-equilibria

Some further remarks on pre-equilibria are relevant to the catalysis of metal ion reactions by anions. In particular, the reaction of Fe^{II} with Fe^{III} may be considered,[6] the reaction being conveniently followed using labelled Fe^{III}

$$Fe^{II} + {}^*Fe^{III} \rightarrow Fe^{III} + {}^*Fe^{II}.$$

Such reactions are generally investigated in aqueous perchloric acid/sodium perchlorate solutions and at constant ionic strength. From a series of experiments at different hydrogen-ion concentrations, the rate is found to have an inverse hydrogen-ion dependence, and with the addition of chloride there is a direct chloride

6. J. Silverman and R. W. Dodson, *J. Phys. Chem.* **56**, 846 (1952).

ion dependence. Under these conditions, then, the reaction can proceed by at least three different paths,

$$Fe^{2+} + Fe^{3+} \xrightarrow{k_1} \quad (1)$$

$$Fe^{2+} + FeOH^{2+} \xrightarrow{k_2} \quad (2)$$

$$Fe^{2+} + FeCl^{2+} \xrightarrow{k_3} \quad (3)$$

where (2) and (3) are faster than (1).

Let us consider, first of all, the acid dependence, where $[H^+]$ is always much in excess of metal-ion concentrations, so that buffering is not necessary. For any one experiment,

$$\text{Rate} = k_{obs}[Fe^{II}][Fe^{III}],$$

and k_{obs} varies as the acid concentration is varied. From (1) and (2) the rate equation may be written

$$\text{Rate} = k_1[Fe^{2+}][Fe^{3+}] + k_2[Fe^{2+}][FeOH^{2+}],$$

and since the pre-equilibrium

$$Fe^{3+} + H_2O \overset{K_H}{\rightleftharpoons} FeOH^{2+} + H^+$$

is fast,

$$[FeOH^{2+}] = \frac{K_H[Fe^{3+}]}{[H^+]},$$

it follows that

$$\text{Rate} = k_1[Fe^{2+}][Fe^{3+}] + \frac{k_2K_H[Fe^{2+}][Fe^{3+}]}{[H^+]}.$$

Providing the amount of Fe^{III} in the form $FeOH^{2+}$ is negligible, this may be written

$$\text{Rate} = k_1[Fe^{II}][Fe^{III}] + \frac{k_2K_H[Fe^{II}][Fe^{III}]}{[H^+]}$$

and

$$k_{obs} = k_1 + k_2K_H/[H^+].$$

From a linear plot of k_{obs} against $1/[H^+]$, values for k_1 and k_2K_H can therefore be obtained.

A more rigorous approach is required in the presence of chloride ions, since there is more extensive complexing of Fe^{3+} to $FeCl^{2+}$ (some $FeCl_2^+$ is also formed but will be ignored for the purpose of this discussion). The fast pre-equilibrium reaction is

$$Fe^{3+} + Cl^- \overset{K}{\rightleftharpoons} FeCl^{2+}$$

where

$$[FeCl^{2+}] = K[Fe^{3+}][Cl^-].\dagger$$

The total concentration of Fe^{III} is therefore

$$[Fe^{III}] = [Fe^{3+}] + [FeOH^{2+}] + [FeCl^{2+}],$$

and substituting for $[FeOH^{2+}]$ and $[FeCl^{2+}]$,

$$[Fe^{III}] = [Fe^{3+}] + [Fe^{3+}]\frac{K_H}{[H^+]} + K[Fe^{3+}][Cl^-],$$

or

$$\frac{[Fe^{III}]}{[Fe^{3+}]} = 1 + \frac{K_H}{[H^+]} + K[Cl^-].$$

Assuming $K_H/[H^+]$ is small compared to $1 + K[Cl^-]$, it follows that

$$[Fe^{3+}] = \frac{[Fe^{III}]}{1 + K[Cl^-]}.$$

Hence the rate equation in k_1, k_2 and k_3 is

$$\text{Rate} = \frac{k_1[Fe^{II}][Fe^{III}] + k_2K_H[Fe^{II}][Fe^{III}]/[H^+] + k_3K[Fe^{II}][Fe^{III}][Cl^-]}{1 + K[Cl^-]}$$

and

$$k_{obs} = \frac{k_1 + k_2K_H/[H^+] + k_3K[Cl^-]}{1 + K[Cl^-]}.$$

---

† We assume, here, that $[Fe^{3+}][Cl^-]$ is proportional to $[FeCl^{2+}]$. In this instance, the formula $FeCl^{2+}$ need not necessarily imply inner-sphere complex ing, but could also refer to outer-sphere complexes of the type $Fe(H_2O)_6, Cl^-$

At constant $[H^+]$,

$$k_{obs} = \frac{A + k_3K[Cl^-]}{1 + K[Cl^-]},$$

where A is a constant

and

$$k_{obs}(1 + K[Cl^-]) = A + k_3K[Cl^-].$$

If, then, K is known, the left-hand side of this equation can be plotted against $[Cl^-]$ and from the gradient k_3 can be obtained.

CHAPTER 4

UNIMOLECULAR GAS-PHASE DECOMPOSITION REACTIONS

It has now been established that a number of decomposition reactions are unimolecular processes, and at high pressures, at least, these give first-order kinetics. Radioactive decay processes are known to have a similar rate dependence, with constants characteristic of the disintegration, but differ from unimolecular reactions in that they are temperature independent, the energy required for the transformation already being present in the nucleus.

Although unimolecular reactions show first-order behaviour at high pressures, they change over to second-order kinetics at lower pressures. The more complex the reactant molecule and the greater its number of degrees of freedom, the lower is the pressure range over which this change takes place. Thus, for organic molecules with more than seven or eight atoms, e.g. ethane and cyclopropane, the change takes place at pressures less than 1 mm Hg. With much simpler inorganic molecules, however, second-order kinetics tend to be observed up to and above atmospheric pressures, and in the decomposition of nitrous oxide, the reaction does not finally become first-order until pressures of around 30 atm.

The change in order of a reaction can be accounted for by considering collisional activation and deactivation of reactant molecules.

Theory of collisional activation (Lindemann's theory)

Consider the decomposition of a molecule A. If an activated molecule A* does not undergo immediate chemical change, then it may become collisionally deactivated,

$$A + A \underset{k_{-1}}{\overset{k_1}{\rightleftharpoons}} A^* + A,$$

$$A^* \xrightarrow{k_2} \text{products.}$$

Assuming stationary-state kinetics for A*,

$$\frac{d[A^*]}{dt} = k_1[A]^2 - k_{-1}[A^*][A] - k_2[A^*] = 0$$

and

$$[A^*] = \frac{k_1[A]^2}{k_{-1}[A] + k_2}.$$

Hence the rate of decomposition can be written

$$\frac{-d[A]}{dt} = k_2[A^*] = \frac{k_1 k_2 [A]^2}{k_{-1}[A] + k_2}.$$

At high pressures such that $k_{-1}[A] \gg k_2$, this reduces to

$$\frac{-d[A]}{dt} = \frac{k_1 k_2 [A]}{k_{-1}} \text{ (i.e. the reaction is first-order)}$$

and at sufficiently low pressures when $k_2 \gg k_{-1}[A]$

$$\frac{-d[A]}{dt} = k_1[A]^2 \text{ (i.e. the reaction is second-order).}$$

The change from first- to second-order kinetics can be indicated graphically by plotting the experimental quantity k, defined by the equation

$$k = \frac{1}{[A]} \frac{d[A]}{dt} = \frac{k_1 k_2 [A]}{k_{-1}[A] + k_2}$$

against [A], inset to Fig. 15. Values of k can be obtained by studying a reaction over its initial stages when the amount of product formed is negligible. At high pressures, k for any one reaction is constant and equal to $k_1 k_2 / k_{-1}$, while, at low pressures, k tends to vary linearly with [A] the gradient being k_1. A better plot for illustrating the effect of increasing complexity of the reactant is

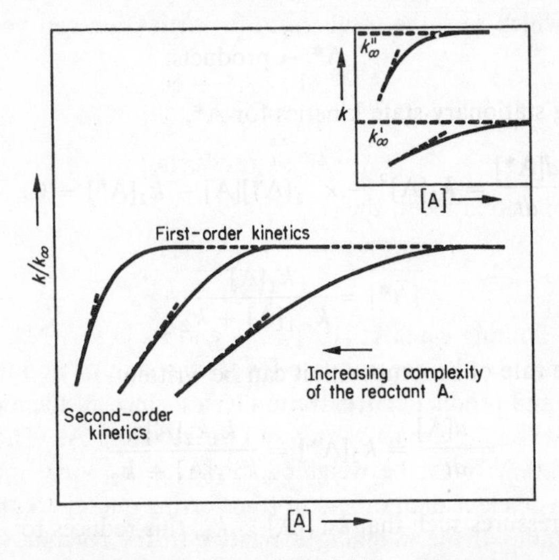

FIG. 15. Diagram showing variation of k/k_∞ with reactant concentration in unimolecular reactions. The curves to the left are for the more complex molecules. The inset defines the quantity k_∞.

that of k/k_∞ against [A], Fig. 15, where k_∞ is the high pressure value of k. Rate constants k_1 obtained from the limiting second-order region are found to exceed calculated $Ze^{-E/RT}$ values where Z is the bimolecular collision frequency and E the experimental activation energy. In other words, the pre-exponential A-factors from experiment are greater than the collision frequency Z (which is of the order 10^{11}–10^{12}), and tend to become increasingly so, as the complexity of the reactant increases. This is because the more degrees of freedom a molecule has, the more chance it has of acquiring the energy E, since this energy may now be distributed amongst several degrees of freedom. Theoretical approaches which may be considered in the calculation of rate constants for unimolecular processes are referred to briefly in a later section.

So far, the effect of a non-reactant gas has been ignored. In many cases, such a gas is present in considerable excess as carrier gas, in which case the mechanism for activation can be written

$$A + M \underset{k_{-1}}{\overset{k_1}{\rightleftharpoons}} A^* + M$$

$$A^* \xrightarrow{k_2} \text{products.}$$

By a similar treatment, then

$$\frac{-d[A]}{dt} = \frac{k_1 k_2 [A][M]}{k_{-1}[M] + k_2},$$

and two limiting cases $k_{-1}[M] \gg k_2$ and $k_2 \gg k_{-1}[M]$, may be considered. In the general case [M] is the sum of non-reactant, reactant and product concentration terms, since all the molecules present will assist in activating and deactivating A. The actual concentrations must be weighted, however, as some molecules are more efficient than others in transferring energy to and from the reactant. If the weighing is relative to the reactant (see, for example, the decomposition of ozone), plots of k against [M] will correspond to those in Fig. 15.

Although collisional activation provides an extremely useful first approach to unimolecular reactions, it is unable to provide a complete explanation of their behaviour. Its inadequacy can, for example, be indicated for systems in which there is a large excess of the non-reactant M. From the treatment given above

$$k = \frac{k_1 k_2 [M]}{k_{-1}[M] + k_2},$$

hence

$$\frac{1}{k} = \frac{k_{-1}}{k_1 k_2} + \frac{1}{k_1} \frac{1}{[M]},$$

and a plot of $1/k$ against $1/[M]$ should therefore be linear. In fact, such plots are curved, as shown in Fig. 16, and as [M] increases, intercepts made by tangents to the experimental curve decrease (i.e. k_{-1}/k_1k_2 values decrease). The explanation put forward for this, is that the specific rate constant k_2 for the de-

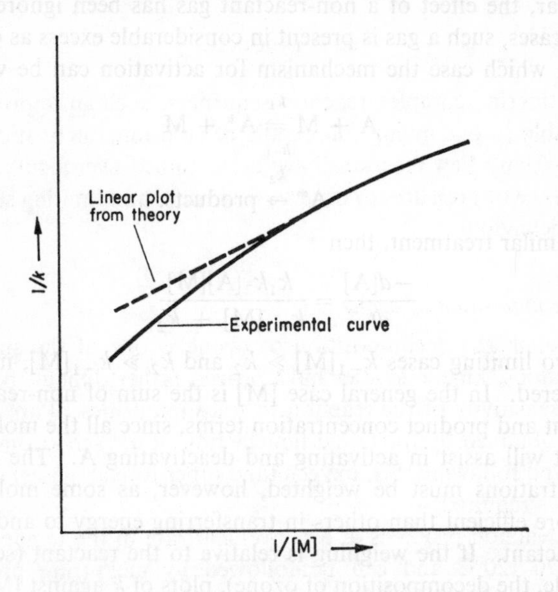

FIG. 16. The experimental curve does not agree with that predicted by collisional activation theory.

composition of activated molecules is not constant but depends on the pressure of the system. At the lower pressures, there is an increasing contribution to the reaction from molecules with smaller transformation probabilities (that is, from molecules which take longer to make up their minds what they will do). As a result of this, the energy distribution curve becomes less populated in the region of still higher energies, and k_2 is smaller at the lower pressures. The graph is particularly important because it indicates a range of values for k_2.

For convenience, a unimolecular reaction studied under essentially second-order conditions may be expressed

$$A + M \rightarrow \text{products},$$

and one studied under first-order conditions,

$$A \rightarrow \text{products}.$$

In considering complex reaction sequences, such an approach is preferable to one giving full details of collisional activation and deactivation. Few reactions have been studied over a sufficiently wide range of pressures to observe first-order and limiting second-order behaviour.

The decomposition of ozone

Although the decomposition of ozone is one of the simplest of reactions, kinetic studies have proved rather difficult because of the sensitivity of the reaction to trace impurities. Pure ozone is best obtained by fractional liquefaction of oxygen–ozone mixtures, and following this, oxygen (and other foreign gases) can be added as required.

The reaction proceeds at a convenient rate in the temperature range 70–120°C and can be followed by measuring pressure changes.[1] In a typical experiment with pure ozone, pressure changes were followed from an initial value of about 100 mm Hg to a final ozone pressure of 10 mm Hg. At higher temperatures, there is an acceleration which can be accounted for in terms of temperature gradients within the system, the reaction being very exothermic,

$$O_3 \rightarrow \tfrac{3}{2}O_2 + 34 \text{ (kcal mole}^{-1}\text{)}.$$

With ozone pressures much above 100 mm, the reaction tends to become explosive. The reaction has also been studied in vessels

1. S. W. Benson and A. E. Axworthy, *J. Chem. Phys.* **26**, 1718 (1957).

of varying size, from 5 to 12 litres and, from such experiments, heterogeneous surface effects would seem to be negligible.

There is no evidence for the direct bimolecular reaction

$$O_3 + O_3 \rightarrow 3\,O_2,$$

and all the existing experimental data can be interpreted in terms of the reaction sequence

$$O_3 + M \underset{k_{-1}}{\overset{k_1}{\rightleftharpoons}} O_2 + O + M,$$

$$O_3 + O \xrightarrow{k_2} 2O_2.$$

Assuming stationary-state kinetics for oxygen atoms,

$$\frac{-d[O]}{dt} = k_1[O_3][M] - k_{-1}[O_2][O][M] - k_2[O_3][O] = 0,$$

and

$$[O] = \frac{k_1[M][O_3]}{k_{-1}[M][O_2] + k_2[O_3]},$$

but the rate of conversion of ozone is given by

$$\frac{-d[O_3]}{dt} = k_2[O_3][O] + k_1[O_3][M] - k_{-1}[O_2][O][M],$$

hence substituting for [O]

$$\frac{-d[O_3]}{dt} = \frac{2k_1k_2[M][O_3]^2}{k_{-1}[M][O_2] + k_2[O_3]}.$$

Using measured gradients G from a plot of $1/[O_3]$ against the time t where

$$G = \frac{d(1/[O_3])}{dt} = \frac{-1}{[O_3]^2}\frac{d[O_3]}{dt},$$

the above equation can be rearranged,

$$\frac{M}{G[O_3]} = \frac{k_{-1}}{2k_1k_2}\frac{[O_2][M]}{[O_3]} + \frac{1}{2k_1}.$$

From a plot of $[M]/G[O_3]$ against $[O_2][M]/[O_3]$, the ratios $k_{-1}/2k_1k_2$ and $1/2k_1$ can be determined.

If the concentration of M at any one time is obtained by direct addition of the ozone and oxygen concentrations, then the above plot does not give straight lines. This is because the efficiency of ozone and oxygen in activating and deactivating ozone molecules are quite different. Only by weighting one of the concentration terms can the experimental points be made to fit a straight line. Thus if

$$[M] = [O_3] + a[O_2]$$

a straight line plot is obtained with $a = 0{\cdot}44$. The effect of different foreign gases can be studied in a similar way, the best fit being given by an expression

$$[M] = [O_3] + 0{\cdot}44[O_2] + 0{\cdot}41[N_2] + 1{\cdot}06[CO_2] + 0{\cdot}34[He].$$

The coefficients in this equation indicate the efficiency of the different gases, relative to ozone, in transferring energy to the internal degrees of freedom of an ozone molecule. The more complex the third-body molecule, the more efficient it is in transferring energy to and from the reactant.

The kinetic experiments described above give k_1 and the ratio k_1k_2/k_{-1}. From these, since the equilibrium constant $K_1 = k_1/k_{-1}$ is known from thermodynamic and spectroscopic data, k_{-1} and k_2 can be determined. By studying the decomposition at different temperatures and plotting $\log_{10}k_1$, $\log_{10}k_{-1}$ and $\log_{10}k_2$ values against $1/T$, these can then be expressed in the form of the Arrhenius equations

$$k_1 = 4{\cdot}61 \times 10^{12} \exp(-24{,}000/RT)\,\text{l mole}^{-1}\,\text{sec}^{-1},$$

$$k_{-1} = 6{\cdot}00 \times 10^{7} \exp(600/RT)\,\text{l}^2\,\text{mole}^{-2}\,\text{sec}^{-1},$$

$$k_2 = 2{\cdot}96 \times 10^{9} \exp(-6{,}000/RT)\,\text{l mole}^{-1}\,\text{sec}^{-1}.$$

The activation energy of 6 kcal mole^{-1} for k_2 is of interest, since this reaction is known to be exothermic by 93 kcal mole^{-1},

$$O_3 + O \rightarrow 2O_2 + 93\ (\text{kcal mole}^{-1}).$$

The oxygen so formed will have 99 kcal mole^{-1} of excess energy to dispose of, and the possibility of energy chains,

$$O_2^* + O_3 \rightarrow 2O_2 + O,$$

has therefore been considered where O_2^* is a high energy oxygen molecule. If such a step were at all significant, however, the overall kinetics would be affected. Since the kinetics are consistent with the mechanism already outlined, it can be concluded that high energy oxygen molecules are quickly deactivated in the normal way.

The negative activation energy of $-0{\cdot}6$ kcal mole^{-1} for the third-order process k_{-1} can be explained by considering the fuller reaction sequence

$$O_3 + M \underset{k_{-A}}{\overset{k_A}{\rightleftharpoons}} O_3^* + M,$$

$$O_3^* \underset{k_{-B}}{\overset{k_B}{\rightleftharpoons}} O_2 + O,$$

$$O_3 + O \xrightarrow{k_C} 2O_2.$$

The rate constants previously defined are given by $k_1 = k_A$, $k_2 = k_C$, while it can be shown that k_{-1} is equivalent to the composite term $k_{-A}k_{-B}/k_B$. Since the apparent activation energy for k_{-1} is given by

$$k_{-1} = A \exp(-E_{-1}/RT)$$

and

$$k_{-A}k_{-B}/k_B = A \exp -(E_{-A} + E_{-B} - E_C)/RT$$

clearly, k_{-1} may be negative if E_C is greater than $E_{-A} + E_{-B}$. A further discussion of negative activation energies, obtained for third-order, but not necessarily single-stage termolecular reactions, is given in the next chapter.

The decomposition of hydrogen peroxide

The decomposition of hydrogen peroxide has been studied at atmospheric and sub-atmospheric pressures, carrier gas included, over a temperature range 240–660°C, using flow methods.[2] Above *ca.* 420°C, the reaction is predominantly homogeneous,

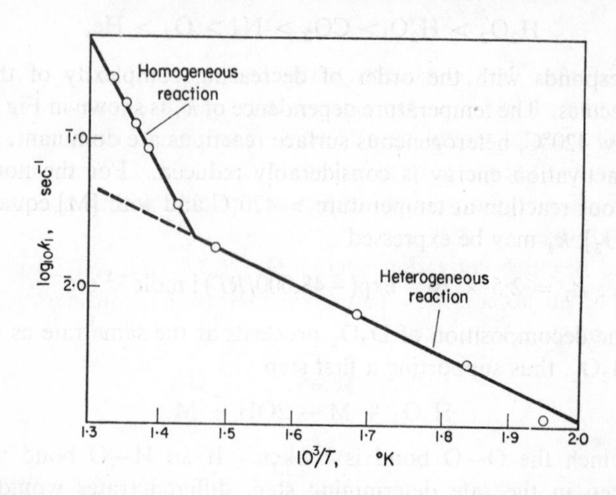

FIG. 17. Arrhenius plot for the decomposition of hydrogen peroxide in a Pyrex vessel. Below *ca.* 420°C heterogeneous surface reactions are predominant. [Reproduced, with permission, from D. E. HOARE, J. B. PROTHEROE and A. D. WALSH, *Trans. Faraday Soc.* **55**, 548 (1959).]

and rates are reproducible in different reactor vessels. The reaction sequence is believed to be

$$H_2O_2 + M \xrightarrow{k_1} 2OH + M$$

$$H_2O_2 + OH \rightarrow H_2O + HO_2$$

$$HO_2 + HO \rightarrow H_2O_2 + O_2,$$

the first step being rate-determining,

$$\frac{-d[H_2O_2]}{dt} = k_1[H_2O_2][M].$$

As in the decomposition of ozone, [M] can be obtained by

2. D. E. HOARE, J. B. PROTHEROE and A. D. WALSH, *Trans. Faraday Soc.* **55**, 548 (1959).

adding weighted concentration terms for all the different species present. The order of efficiency of third-bodies,

$$H_2O_2 > H_2O > CO_2 > N_2 > O_2 > He,$$

corresponds with the order of decreasing complexity of these molecules. The temperature dependence of k_1 is shown in Fig. 17. Below 420°C, heterogeneous surface reactions are dominant, and the activation energy is considerably reduced. For the homogeneous reaction at temperature > 420°C and with [M] equal to $[H_2O_2]$, k_1 may be expressed

$$k_1 = 2{\cdot}5 \times 10^{15} \exp(-48{,}000/RT)\ \text{l mole}^{-1}\ \text{sec}^{-1}$$

The decomposition of D_2O_2 proceeds at the same rate as that of H_2O_2, thus supporting a first step

$$H_2O_2 + M \rightarrow 2OH + M$$

in which the O—O bond is broken. If an H—O bond were broken in the rate determining step, different rates would be expected, since the H—O and D—O bond energies are different.

The decomposition of dinitrogen tetroxide

Shock tube techniques have been used to investigate the decomposition of dinitrogen tetroxide to an equilibrium mixture,[3]

$$N_2O_4 \rightleftharpoons 2NO_2.$$

The reaction is conveniently followed by measuring the absorption of the 400 mμ Hg line by NO_2. In a typical experiment, nitrogen at one atmosphere and containing 1% N_2O_4 was shocked, the driver gas being nitrogen at 2 atm, and the temperature of the shocked gas increased by some 25°C. Since the rate of dissociation is too fast to measure above 30°C, the whole of the shock tube had to be pre-cooled to temperatures down to −35°C. Even so, over the temperature range −20–+20°C, reactions were complete within 10^{-3} sec and electronic means were required to record the colorimetric changes.

3. T. Carrington and N. Davidson, *J. Phys. Chem.* **57**, 418 (1953).

For N_2O_4 plus carrier-gas pressures of around one atmosphere, the decomposition follows second-order kinetics,

$$N_2O_4 + M \rightarrow 2NO_2 + M.$$

At pressures of around three atmospheres, the kinetics approach first-order behaviour, and from a plot of the kind shown in Fig. 15, the first-order constant k_∞ can be estimated. From an Arrhenius plot of k_∞ against $1/T$,

$$k_\infty = 10^{16} \exp(-13{,}100/RT) \text{ sec}^{-1}.$$

The decomposition of dinitrogen pentoxide

Dinitrogen pentoxide can be prepared by dehydrating nitric acid with phosphorous pentoxide. Although, in the solid phase, " N_2O_5 " is $NO_2^+NO_3^-$, in the gas phase (b.p. 47°C) it probably has a planar structure

$$\begin{matrix} O & & & & O \\ & \rangle N - & O & - N \langle & \\ O & & & & O \end{matrix}$$

The gas-phase decomposition can be studied over a temperature range 0–120°C either by following pressure changes or spectrophotometrically, since NO_2 is coloured. The overall equation is

$$\begin{matrix} 2N_2O_5 \rightarrow & 2N_2O_4 + O_2, \\ & \Downarrow\!\Uparrow \\ & 2NO_2 \end{matrix}$$

where the equilibrium between N_2O_4 and NO_2 can be allowed for, since it is fast, and the equilibrium constant is known.

The reaction obeys the first-order rate equation

$$-d[N_2O_5]/dt = k_{obs}[N_2O_5],$$

over a wide range of conditions. Although a number of reaction paths are possible, it is now well established[4] that the mechanism is, in fact,

$$N_2O_5 \underset{k_{-1}}{\overset{k_1}{\rightleftharpoons}} NO_3 + NO_2$$

$$NO_3 + NO_2 \xrightarrow{k_2} NO + O_2 + NO_2$$

$$NO_3 + NO \xrightarrow{k_3} 2NO_2.$$

The experimental rate equation can be derived as follows. First, assuming stationary-state kinetics for the NO_3 radical,

$$\frac{d[NO_3]}{dt} = k_1[N_2O_5) - k_{-1}[NO_3][NO_2] - k_2[NO_3][NO_2] - k_3[NO_3][NO] = 0,$$

and

$$[NO_3] = \frac{k_1[N_2O_5]}{k_{-1}[NO_2] + k_2[NO_2] + k_3[NO]}$$

A similar approximation can be made for NO, since it also behaves as a highly reactive intermediate, thus

$$\frac{d[NO]}{dt} = k_2[NO_3][NO_2] - k_3[NO][NO_3] = 0,$$

and

$$[NO] = \frac{k_2[NO_2]}{k_3}.$$

But the overall rate is given by

$$\frac{-d[N_2O_5)}{dt} = 2k_2[NO_3][NO_2],$$

and substituting for $[NO_3]$ and $[NO]$,

$$\frac{d[N_2O_5]}{dt} = \frac{2k_1k_2[N_2O_5]}{k_{-1} + 2k_2}.$$

4. R. A. Ogg, Jr., *J. Chem. Phys.* **15**, 337 (1947).

This is of the same form as the experimental rate equation, where k_{obs} can be identified as the composite term

$$\frac{2k_1k_2}{k_{-1}+2k_2}.$$

Three other first steps are possible and should be mentioned. These are

$$N_2O_5 \rightarrow N_2O_4 + O,$$

$$N_2O_5 \rightarrow NO_2 + NO + O_2,$$

$$N_2O_5 \rightarrow N_2O_3 + O_2.$$

In the first, the N_2O_4 is an isomeric form of normal dinitrogen tetroxide, i.e. it is $ONONO_2$ and not O_2NNO_2. Such a reaction is unlikely since it would require an exceptionally high activation energy of around 60 kcal mole^{-1}. The second is unlikely since it requires the simultaneous breaking of two bonds and the formation of three molecules from one. In the third, moreover, oxygen has two unpaired electrons, while both N_2O_5 and N_2O_3 are diamagnetic. Reactions of this type involving electron spin changes are improbable and would be necessarily slow.

For a long time k_{obs} was thought to be equivalent to k_1, subsequent reactions being fast and not, therefore, rate determining. Evidence for the mechanism already given was obtained from exchange experiments between $^{15}N_2O_5$ and NO_2.[5] If the first step is the rapid equilibrium

$$N_2O_5 \rightleftharpoons NO_3 + NO_2,$$

then the exchange of ^{15}N between N_2O_5 and NO_2 should be measurable. The exchange was found to be much faster than the overall decomposition, thus supporting a reversible first step involving NO_2. There is, furthermore, spectroscopic evidence for the existence of the NO_3 radical in the $N_2O_5 + O_3$ system.

5. R. A. Ogg, W. S. Richardson and M. K. Wilson, *J. Chem. Phys.* **18**, 573 (1950).

Further confirmation of the above mechanism was obtained from a study of the reaction of N_2O_5 with nitric oxide.[6] With an excess of nitric oxide the overall equation is

$$N_2O_5 + NO \rightarrow 3NO_2,$$

the reaction being many times faster than the N_2O_5 decomposition. The reaction shows a first-order dependence on N_2O_5 only, which can be accounted for by considering a mechanism

$$N_2O_5 \underset{k_{-1}}{\overset{k_1}{\rightleftharpoons}} NO_3 + NO_2,$$

$$NO_3 + NO \xrightarrow{k_3} 2NO_2,$$

where the rate constants k_1, k_{-1} and k_3 are as defined previously. Making the stationary-state approximation for $[NO_3]$,

$$[NO_3] = \frac{k_1[N_2O_5]}{k_{-1}[NO_2] + k_3[NO]},$$

a rate equation

$$\frac{-d[N_2O_5]}{dt} = \frac{k_1k_3[N_2O_5][NO]}{k_{-1}[NO_2] + k_3[NO]}$$

can be derived. If now $k_3[NO] \gg k_{-1}[NO_2]$, this reduces to a simple first-order equation

$$\frac{-d[N_2O_5]}{dt} = k_1[N_2O_5],$$

which is in agreement with experiment. In this way, then, k_1 can be measured directly, while in the N_2O_5 decomposition a composite constant only is obtained. The reaction between N_2O_5 and NO has been studied over a wide (10^5) range of pressures,[7] and rate constants near the high and low pressure limits have been

6. J. H. Smith and F. Daniels, *J. Am. Chem. Soc.* **69**, 1735 (1947).
7. R. L. Mills and H. S. Johnston, *J. Am. Chem. Soc.* **73**, 938 (1951).

observed. From the temperature dependence of the high pressure first-order constant k_1,

$$k_1 = 6 \times 10^{14} \exp(-21{,}000/RT)\ \text{sec}^{-1}.$$

The decomposition of nitryl chloride

The thermal decomposition of nitryl chloride,[8]

$$2ClNO_2 \rightarrow Cl_2 + 2NO_2,$$

has been investigated at low pressures in a 50 l. Pyrex reaction vessel. For temperatures in the 180–250°C range, results are in agreement with a mechanism

$$ClNO_2 + M \xrightarrow{k_1} Cl + NO_2 + M$$

$$ClNO_2 + Cl \xrightarrow{\text{fast}} Cl_2 + NO_2,$$

and the rate of decomposition is therefore

$$\frac{-d[ClNO_2]}{dt} = k_1[ClNO_2][M].$$

With $[ClNO_2]$ for [M], k_1 can be expressed

$$k_1 = 6{\cdot}3 \times 10^{13} \exp(-27{,}500/RT)\ \text{l mole}^{-1}\ \text{sec}^{-1}.$$

The decomposition of nitrous oxide

The decomposition of the linear nitrous oxide molecule, N—N—O, to nitrogen and oxygen has been studied at temperatures in the 600°C region.[9] At pressures exceeding 30 atm, the reaction is first order in nitrous oxide, thus excluding the direct bimolecular process

$$N_2O + N_2O \rightarrow 2N_2 + O_2.$$

---

8. H. F. Cordes and H. S. Johnston, *J. Am. Chem. Soc.* **76**, 4264 (1954).
9. C. N. Hinshelwood and F. J. Lindars, *Proc. Roy. Soc.* **A231**, 178 (1955).

Of the two possible first steps

$$N_2O \rightleftharpoons N_2 + O$$

and

$$N_2O \rightleftharpoons N + NO,$$

the latter can be excluded since there is no exchange of nitrogen-atoms during the decomposition of nitrous oxide doubly labelled

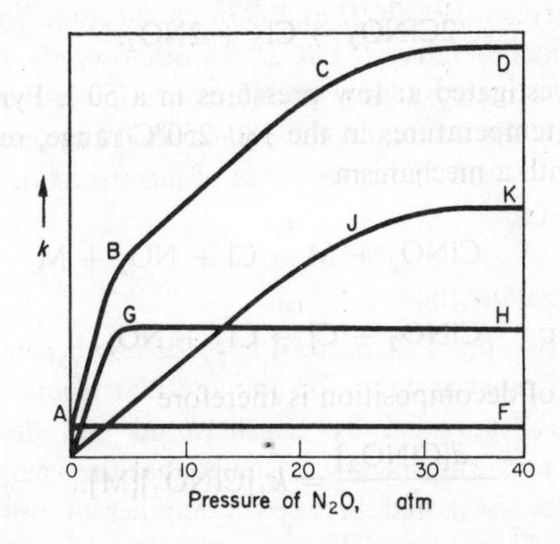

FIG. 18. The experimental curve OABCD for the decomposition of nitrous oxide gives three quasi-unimolecular curves OEF, OGH, and OJK. [Reproduced with permission, E. HUNTER, *Proc. Roy. Soc.* **A144**, 386 (1934).]

with nitrogen-15. The formation of small amounts of nitric oxide along with the nitrogen and oxygen can be accounted for by a mechanism

$$N_2O \rightleftharpoons N_2 + O$$

$$N_2O + O \rightarrow N_2 + O_2$$

$$N_2O + O \rightarrow 2NO$$

$$O + O + M \rightarrow O_2 + M.$$

At low pressures, the reaction shows second-order behaviour

but not quite in the expected manner. Thus if the quantity k defined by an equation

$$k = \frac{1}{[N_2O]} \frac{d[N_2O]}{dt}$$

is plotted against the pressure of N_2O, the form of the curve is not as in other unimolecular processes (Fig. 18). Using methods of curve analysis, it has been shown that three separate unimolecular curves *OEF*, *OGH* and *OJK* can account for the shape of the experimental curve *OABCD*. In other words, there would appear to be a superposition of different unimolecular processes in this system. To account for these curves it has been suggested that unstable $^3\Sigma$ and $^3\pi$ triplet states of nitrous oxide are formed as intermediates.[10]

Theoretical considerations

There are two main theoretical approaches to consider in the calculation of rate constants for unimolecular reactions.[11] The first originally suggested by Hinshelwood and developed by Kassel, Rice and Ramsperger, assumes that the energy required for dissociation is shared between a number of normal vibrational modes between which it can flow freely. The changes which occur are best considered in terms of the reaction sequence

$$A + A \underset{k_{-1}}{\overset{k_1}{\rightleftharpoons}} A^* + A$$

$$A^* \overset{k_2}{\rightarrow} A^\ddagger$$

$$A^\ddagger \overset{k'_2}{\rightarrow} \text{products},$$

where $A^\ddagger$ is an activated molecule which can undergo decomposition and A^* is an energized molecule possessing activation energy E distributed at random. The latter must undergo

10. B. G. Reuben and J. W. Linnett, *Trans. Faraday Soc.* **55,** 1543 (1959).
11. N. B. Slater, *Theory of Unimolecular Reactions*, Methuen, London, 1959.

vibrational changes before it can become an activated molecule, i.e. energy must become localized in one bond. The number of vibrational modes which are active in the conversion of an energized to an activated molecule has to be determined by a trial and error procedure. About half the total number are generally required to give reasonable agreement with experimental results, although no reasonable interpretation of this has yet been given. The rate constant k'_2 is of the order of a vibration frequency, i.e. 10^{13}–10^{14} sec^{-1}.

The second approach by Slater assumes that the flow of energy between different normal modes is a relatively slow process, the decomposition of an energized molecule occurring only when certain modes are sufficiently in phase that the extension of a critical coordinate can take place. Both methods have had a certain amount of success, but neither is completely satisfactory. With hydrogen peroxide, for example,[12] the second-order rate calculated using Slater's theory is too low compared with the experimental value, while the HKRR theory gives good agreement if five or six normal modes are considered active. For ozone, on the other hand,[13] the rate of energization obtained using Slater's theory is in good agreement with experiment, while the rate obtained using the HKRR approach is too high by a factor of three.

12. E. K. Gill and K. J. Laidler, *Proc. Roy. Soc.* **A251**, 66 (1959).
13. E. K. Gill and K. J. Laidler, *Trans. Faraday Soc.* **55**, 753 (1959).

CHAPTER 5

THIRD-ORDER GAS-PHASE REACTIONS

A NUMBER of reactions have been shown to obey third-order kinetics over a wide range of experimental conditions. Such reactions fall into two groups (*a*) the reactions of nitric oxide,

$$2NO + X_2 \rightarrow 2NOX,$$

where X_2 may be O_2, Cl_2, Br_2 and possibly H_2, and (*b*) atom-recombination reactions, e.g.

$$I + I + M \rightarrow I_2 + M,$$
$$Br + Br + M \rightarrow Br_2 + M.$$

In the latter, M is a non-reactant third-body or chaperon, its function being to remove the excess kinetic energy of the two atoms, and thus allow a stable diatomic molecule to be formed.

According to one estimate,[1] the number of ternary collisions in a gas at atmospheric pressure is about 10^2 times smaller than the number of binary collisions. The probability of a transition complex being formed as a result of a ternary collision is small, however, since a considerable rearrangement of translational and rotational energy is required. Thus the reactants have altogether nine translational degrees of freedom, while the transition complex has only three. Because of this, it is generally assumed that third-order reactions take place in stages. Possible mechanisms for a reaction of the general type 2A + B are

$$A + B \overset{\text{fast}}{\rightleftharpoons} AB$$
$$AB + A \rightarrow \text{products},$$

1. S. W. BENSON, *Fundamentals of Chemical Kinetics*, pp. 305–8, McGraw-Hill, New York, 1960.

and/or

$$A + A \overset{\text{fast}}{\rightleftharpoons} A_2$$

$$A_2 + B \rightarrow \text{products,}$$

equilibrium concentrations of the intermediates AB and A_2 being small. As long as the first step is much faster than the subsequent reaction, so that an equilibrium concentration of the intermediate is maintained, both will give third-order rate equations of the required type

$$\frac{-d[A]}{dt} = k[A]^2[B].$$

If intermediates are, in fact, formed, a certain amount of kinetic evidence (by having one of the reactants present in high concentration, for details see later) or more direct evidence (from, say, spectroscopic measurements) might have been expected. Since evidence of this kind has not been obtained (e.g. third-order kinetics persist over a wide range of conditions) direct confirmation of one or both of the above mechanisms is lacking.

In principle, the initial reaction in each of the above mechanisms need be no more than a collision of long duration, i.e. a "sticky collision". Some degree of chemical bonding seems likely, however, since, in this way, the negative activation energies which are obtained from third-order rate constants can be explained (Table 4). Thus, for a two-stage process of the above

TABLE 4

KINETIC DATA FOR SOME THIRD-ORDER REACTIONS AT 25°C

Reaction	k (l^2 mole^{-2} sec^{-1})	log A (l^2 mole^{-2} sec^{-1})	E (kcal mole^{-1})
$2NO + O_2 \rightarrow 2NO_2$	$7{\cdot}1 \times 10^3$	3·0	−1·1
$2NO + Cl_2 \rightarrow 2ClNO$	21·0	3·7	3·7
$2NO + Br_2 \rightarrow 2BrNO$	$3{\cdot}1 \times 10^3$	4·6	−1·5
$2I + M \rightarrow I_2 + M (M = Ar)$	6×10^7	8·8	−1·3
$2Br + M \rightarrow Br_2 + M (M = Ar)$	2×10^9	8·3	−1·4

type, the observed activation energy is the sum of an enthalpy term for the equilibrium (which may be negative) and an activation energy.

From collision theory, the frequency factors for these reactions should be in the region 10^9–10^{10} l^2 $mole^{-2}$ sec^{-1}. This is in general agreement with experiment since *A* factors of this order of magnitude are observed for the faster atom-recombination reactions. For the nitric oxide reactions, *A* factors are considerably less however. In these reactions, orientation of reactants to one another will obviously be important, and steric *P*-factors in the region 10^{-6}–10^{-7} are not altogether unexpected.

The reaction between nitric oxide and oxygen

Over a fairly wide range of conditions the reaction

$$2NO + O_2 \rightarrow 2NO_2,$$

can be followed manometrically. At high reactant concentrations when the reaction is much faster, spectrophotometric methods are generally more convenient. The rapid equilibrium between NO_2 and N_2O_4 must be taken into account, especially at lower temperatures, and the reaction then obeys a rate equation

$$\frac{-d[NO]}{dt} = -2\frac{d[O_2]}{dt} = k[NO]^2[O_2].$$

Above 360°C, the back reaction is significant and must be allowed for. Thus

$$\frac{-d[NO]}{dt} = k[NO]^2[O_2] - k'[NO_2]^2,$$

where the rate constant k' for the back reaction may be expressed in terms of the rate constant k for the forward reaction and the equilibrium constant K, i.e. $k' = k/K$. When the reactant concentrations are not equal, which, in practice, is difficult to achieve, the integrated form of the rate equation is complex and the graphical method in which gradients $d[NO]/dt$ are measured is often preferred. Some typical results are listed in Table 5.[2]

TABLE 5

DATA FOR A SERIES OF EXPERIMENTS ON THE REACTION BETWEEN NITRIC OXIDE AND OXYGEN AT 25°C

Ratio $[NO]/[O_2]$	Range of initial NO pressures (mm)	Number of runs	Average rate constant (l^2 mole^{-2} sec^{-1})
10	10·6–340	28	$6{\cdot}82 \times 10^3$
1	18·4–130	15	$7{\cdot}73 \times 10^3$
0·1	8·1–43·0	20	$7{\cdot}10 \times 10^3$
		Mean value	$7{\cdot}13 \times 10^3$

Two-stage mechanisms require the formation of either $(NO)_2$ or NO_3 (or both). They are

$$2NO \underset{k_{-1}}{\overset{k_1}{\rightleftharpoons}} (NO)_2 \qquad (1)$$

$$(NO)_2 + O_2 \xrightarrow{k_2} 2NO_2, \qquad (2)$$

and

$$NO + O_2 \underset{k_{-3}}{\overset{k_3}{\rightleftharpoons}} NO_3 \qquad (3)$$

$$NO_3 + NO \xrightarrow{k_4} 2NO_2. \qquad (4)$$

In the first, equations (1) and (2), assuming stationary-state kinetics for $(NO)_2$,

$$\frac{-d[(NO)_2]}{dt} = k_1[NO]^2 - k_{-1}[(NO)_2] - k_2[(NO)_2][O_2] = 0,$$

and

$$[(NO)_2] = \frac{k_1[NO]^2}{k_{-1} + k_2[O_2]}.$$

2. H. S. JOHNSTON and L. W. SLENTZ, *J. Am. Chem. Soc.* **73**, 2948 (1951).

The rate equation is therefore

$$\frac{-d[\mathrm{NO}]}{dt} = \frac{k_1k_2[\mathrm{NO}]^2[\mathrm{O}_2]}{k_{-1} + k_2[\mathrm{O}_2]},$$

which is third-order if $k_{-1} \gg k_2[O_2]$, that is, if the initial equilibrium is fast. Under these conditions the final rate equation becomes

$$\frac{-d[\mathrm{NO}]}{dt} = K_1k_2[\mathrm{NO}]^2[\mathrm{O}_2].$$

Using a similar procedure, equations (3) and (4) give a rate equation

$$\frac{-d[\mathrm{NO}]}{dt} = \frac{k_3k_4[\mathrm{NO}]^2[\mathrm{O}_2]}{k_{-3} + k_4[\mathrm{NO}]},$$

and with $k_{-3} \gg k_4[NO]$ this approximates to

$$-d[\mathrm{NO}]/dt = K_3k_4[\mathrm{NO}]^2[\mathrm{O}_2].$$

Evidence supporting one or other of these mechanisms is difficult to obtain, and third-order behaviour is generally observed.

Two-stage mechanisms are particularly attractive in that they give a reasonable explanation of the negative activation energies. Since the third-order rate constant is a composite term the activation energy E_{obs} may be expressed

$$E_{obs} = \Delta H_1 + E_2$$

or

$$E_{obs} = \Delta H_3 + E_4,$$

where ΔH_1 and ΔH_3 are enthalpies (which are negative for exothermic reactions) and E_2 and E_4 are activation energies. If ΔH_1 is more negative than E_2, or ΔH_3 is more negative than E_4, then the experimental quantity E_{obs} will be negative. The activation energies for E_2 and E_4 are probably small and close to zero.

Of the two-stage mechanisms for this and other nitric oxide reactions, the one involving $(NO)_2$ would seem the more likely, since nitric oxide has a single unpaired electron. Nitric oxide is known to dimerize in solution, the heat of association being

$\sim$4 kcal mole^{-1}, and there is some spectroscopic evidence for its formation in the gas phase.[c3a] Even so, it does not necessarily follow that $(NO)_2$ is an intermediate in the reaction. The only sure way of confirming either mechanism is to demonstrate the applicability of the full rate equation

$$\frac{-d[NO]}{dt} = \frac{k_1 k_2 [NO]^2 [O_2]}{k_{-1} + k_2 [O_2]}.$$

Some evidence has now been obtained for this rate law.[3b]

If NO_3 is formed as an intermediate it can be reasoned that this must have a peroxo structure O—O—N—O rather than the symmetrical nitrate form. This follows from a kinetic study of the NO + O_2 system in the presence of a large excess of nitrogen dioxide ($[NO_2]/[NO] \sim 10^3$).[3c] Under these conditions the kinetics are consistent with reactions

$$NO + NO_2 + O_2 \underset{k_{-5}}{\overset{k_5}{\rightleftharpoons}} NO_3 + NO_2 \qquad (5)$$

$$NO_3 + NO \rightarrow 2NO_2, \qquad (6)$$

together with the third-order reaction

$$2NO + O_2 \rightarrow 2NO_2. \qquad (7)$$

If, then, NO_3 is an intermediate in (7) it cannot have the same form as the NO_3 in (5). If it did, the rate equation would be of a different form from that actually observed. Since the rate constant k_{-5} is in agreement with values obtained from the study of N_2O_5 systems, and it can be assumed that NO_3 from the latter is of the nitrate form, it follows that any NO_3 formed in (7) must be of the peroxo kind.

The reactions between NO and Cl_2, and NO and Br_2

Both these reactions can be followed manometrically. The first

$$2NO + Cl_2 \rightarrow 2ClNO,$$

3. (a) See for example report in *Chem. and Eng. News*, **47**, 42 (1969); (b) M. Šok, *Nature*, **209**, 706 (1966), and (c) J. D. Ray and R. A. Ogg, *J. Chem. Phys.* **26**, 984 (1957).

has been studied at temperatures up to 300°C,[4] and the second,

$$2NO + Br_2 \rightarrow 2BrNO,$$

over a more limited range, −8 to 15°C.[5] At all these temperatures, an allowance has to be made for the back reactions, the forward reactions conforming to third-order rate expressions,

$$\frac{d[ClNO]}{dt} = k'[NO]^2[Cl_2]$$

and

$$\frac{d[BrNO]}{dt} = k''[NO]^2[Br_2].$$

The curvature of the Arrhenius plot of k' at the higher temperatures (Fig. 19) can be attributed to radical reactions. These are probably initiated by the decomposition

$$ClNO \rightarrow Cl + NO.$$

To account for the third-order kinetics, two-stage bimolecular processes may be considered as before, possible intermediates being either $(NO)_2$ or Cl_2NO and Br_2NO.

The reaction between nitric oxide and hydrogen

The reaction of hydrogen with nitric oxide proceeds at a measurable rate above 700°C, the final products being given by an equation

$$2NO + 2H_2 \rightarrow N_2 + 2H_2O.$$

At 801°C, the rate is proportional to $[NO]^2$ and $[H_2]$ over a fairly wide but limited range of concentrations,[6] and the rate

---

4. I. Welinsky and H. A. Taylor, *J. Chem. Phys.* **6**, 466 (1938).
5. W. Krauss, *Zeit. Physik Chem. (Leipzig)*, A**175**, 295 (1936).
6. C. N. Hinshelwood and J. W. Mitchell, *J. Chem. Soc.* 378 (1936).

determining step or steps are thought to involve one hydrogen and two nitric oxide molecules. Possible reactions are

$$2NO + H_2 \rightarrow N_2 + H_2O_2,$$

$$2NO + H_2 \rightarrow N_2O + H_2O,$$

$$2NO + H_2 \rightarrow 2HNO,$$

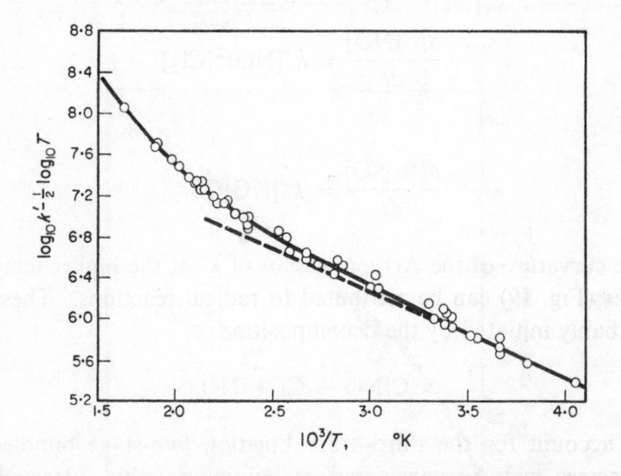

FIG. 19. Activation energy plot for the reaction of chlorine with nitric oxide. (The scale of the ordinate axis is in units of cc² mole⁻² sec⁻¹.) The plot still shows curvature even though the small temperature dependence of the pre-exponential term is taken into account. [Reproduced with permission from I. WELINSKY and H. A. TAYLOR, *J. Chem. Phys.* 6, 466 (1938).]

where the intermediates so formed react rapidly with further quantities of hydrogen to give the final products. Evidence for the existence of HNO radicals has been obtained by reacting hydrogen atoms with nitric oxide,[7] a yellow deposit of $(HNO)_n$ being obtained at liquid air temperatures.

With both reactants at pressures of 300 mm or above, graphs of the initial rate against pressure of hydrogen, at a constant nitric oxide pressure, are linear over a wide range. When the linear

7. H. A. TAYLOR and H. A. TANFORD, *J. Chem. Phys.* 6, 466 (1938).

section is produced back, however, it does not pass through the origin, and, at lower hydrogen pressures, instead of continuing linearly to a finite intercept, it bends round and passes through the origin (Fig. 20). The shape of the curve can be accounted for by a rate equation

$$\frac{-d[H_2]}{dt} = k_1[NO]^2[H_2] + \frac{k_2[NO]^2[H_2]}{1 + a[NO]} + \frac{k_3[NO]^2[H_2]}{1 + b[H_2]},$$

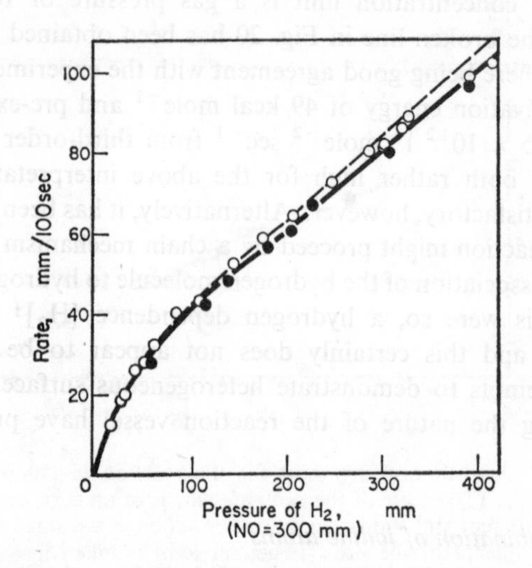

FIG. 20. The reaction between nitric oxide and hydrogen at 801°C. Open circles are for experiments in an unpacked, and shaded circles for experiments in a packed reactor. The broken line is for values calculated from the formula given in the text. [Reproduced, with permission, from C. N. HINSHELWOOD and J. W. MITCHELL, *J. Chem. Soc.* 378 (1936).]

where a and b are constants. The latter is obtained by considering parallel reaction paths involving

(1) ternary collisions,

(2) the formation of an intermediate H_2NO

and (3) the formation of an intermediate $(NO)_2$.

Initial rates measured as carefully as possible by drawing tangents to a concentration–time curve, and were found to fit an equation

$$\frac{-d[H_2]}{dt}\ (\text{mm}/100\ \text{sec}) = 1{\cdot}0[NO]^2[H_2] + \frac{30[NO]^2[H_2]}{1 + 8{\cdot}0[NO]} + \frac{11[NO]^2[H_2]}{1 + 3{\cdot}4[H_2]},$$

where the concentration unit is a gas pressure of 100 mm at 801°C. The broken line in Fig. 20 has been obtained using this formula, there being good agreement with the experimental plot.

The activation energy of 49 kcal mole^{-1} and pre-exponential factor of 5×10^{12} l^2 mole^{-2} sec^{-1} from third-order rate constants are both rather high for the above interpretation to be entirely satisfactory, however. Alternatively, it has been suggested that the reaction might proceed by a chain mechanism following thermal dissociation of the hydrogen molecule to hydrogen atoms, but, if this were so, a hydrogen dependence $[H_2]^{\frac{1}{2}}$ would be expected, and this certainly does not appear to be the case. Other attempts to demonstrate heterogeneous surface reactions by varying the nature of the reaction vessel have proved unsuccessful.

The recombination of iodine atoms

Flash photolysis techniques can be used to obtain high concentrations of iodine atoms,[8] and are generally preferred to shock-tube methods which give less precise results. In a typical series of experiments, argon at 100 mm pressure and containing up to 0·13 mm of iodine is flashed (duration of flash $\sim 10^{-5}$ sec), some 20% dissociation occurring with a flash of sufficiently high intensity. The subsequent recombination of iodine atoms can be followed spectrophotometrically, the iodine molecule but not the iodine atom absorbing in the visible. A fixed wavelength is

8. G. Porter and J. A. Smith, *Proc. Roy. Soc.* **A261**, 28 (1961); R. Engelman and N. Davidson, *J. Am. Chem. Soc.* **82**, 4770 (1960).

used, the 520 mμ iodine maximum being most convenient. After passing through the reactor, the light is fed to a photo-multiplier and the response shown on an oscillograph. The time-base of the

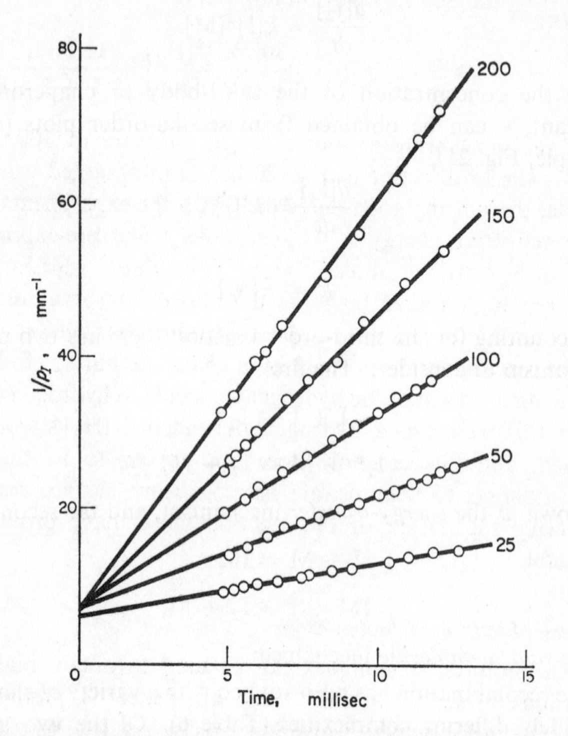

FIG. 21. Second-order plots for the iodine atom recombination in the presence of gas at pressures 25–200 mm. [Reproduced, with permission, from M. I. CHRISTIE, R. G. W. NORRISH and G. PORTER, *Proc. Roy. Soc.* A**216**, 158 (1953).]

latter is triggered by the initial flash and the concentration–time trace for the whole of the reaction period, generally a few milliseconds, is photographed. This shows, first of all, a sharp decrease in iodine concentration, as a result of the flash, and is followed by a somewhat slower increase as the iodine is re-formed. Concentration–time data for the recombination obeys the third-order rate equation

$$\frac{d[I_2]}{dt} = k[I]^2[M].$$

Since the concentration of the third-body or chaperon M is constant, k can be obtained from second-order plots (see for example, Fig. 21),

$$\frac{d[I_2]}{dt} = k'[I]^2,$$

where

$$k' = k[M].$$

In accounting for the third-order kinetics, there are two possible mechanism to consider. The first,

$$I + I \rightleftharpoons I_2^*$$

$$I_2^* + M \rightarrow I_2 + M,$$

is known as the *energy-transfer* mechanism, and the second,

$$I + M \rightleftharpoons IM$$

$$IM + I \rightarrow I_2 + M,$$

as the *radical–molecule* mechanism.

The recombination has been studied with a variety of chaperons of widely differing complexities (Table 6). Of the two possible reaction sequences the *radical–molecule* mechanism is thought to predominate. Thus, third-order rate constants vary over a far greater range than can be understood in terms of the collision diameters or energy-transfer efficiencies of the different chaperons (as would be required in the energy-transfer mechanism). To account for the negative activation energies, specific bonding is necessary and it has been suggested that charge-transfer type complexes are formed. In solution media, with the chaperon as solvent, spectral evidence has been obtained for a number of IM

TABLE 6

KINETIC DATA FOR THE RECOMBINATION OF IODINE ATOMS WITH A VARIETY OF CHAPERONS

M	k (l^2 mole^{-2} sec^{-1}) $\times 10^{-9}$	E (kcal mole^{-1})
He	1·5	−0·53
Ne	1·9	−1·5
Ar	3·0	−1·27
H_2	5·7	−1·22
O_2	6·8	−1·5
CO_2	13·4	−1·75
C_4H_{10}	36	−1·65
C_6H_6	92	−1·83
CH_3I	160	−2·55
$C_6H_5CH_3$	194	−2·7
C_2H_5I	262	−2·4
$C_6H_3(CH_3)_3$	405	−4·1
I_2	$1·6 \times 10^3$	−4·4
NO	3×10^4	−2·2

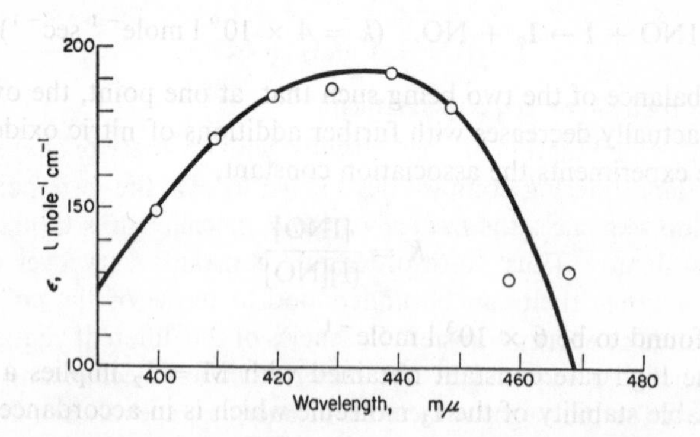

FIG. 22. The absorption spectrum of the intermediate INO. [Reproduced, with permission, from G. PORTER, Z. G. SZABO and M. G. TOWNSEND, *Proc. Roy. Soc.* A270, 493 (1962).]

complexes, and there are good reasons for supposing that these will have bond energies of the required order of magnitude. Whether the reactions are exclusively by the radical–molecule mechanism or not is more difficult to answer. With the inert gases, at least, IM complexes would seem less likely and the energy-transfer mechanism may make some contribution.

In one case only, with nitric oxide as chaperon, has direct evidence (kinetic and spectroscopic) been obtained for the intermediate IM.[9] By flashing a mixture containing 3·6 mm of iodine (which is much higher than is generally used) 250 mm of nitric oxide and a further 100 mm of neon as diluent gas, a transient absorption with a maximum at 430 mμ is observed (Fig. 22). The latter has been attributed to the intermediate formation of the otherwise unknown compound INO (i.e. the analogue of BrNO and ClNO). The stability of INO is such that third-order rate equations are obeyed only at nitric oxide pressure of up to 0·4 mm. At higher pressures, with increasing quantities of INO, the reaction

$$\mathrm{INO} + \mathrm{INO} \rightarrow \mathrm{I_2} + 2\mathrm{NO}, \ (k < 4 \times 10^7 \ \mathrm{l\ mole^{-1}\ sec^{-1}})$$

competes with the normal reaction path,

$$\mathrm{INO} + \mathrm{I} \rightarrow \mathrm{I_2} + \mathrm{NO}, \quad (k = 4 \times 10^9 \ \mathrm{l\ mole^{-1}\ sec^{-1}})$$

The balance of the two being such that, at one point, the overall rate actually decreases with further additions of nitric oxide. In these experiments the association constant,

$$K = \frac{[\mathrm{INO}]}{[\mathrm{I}][\mathrm{NO}]},$$

was found to be 6×10^3 l mole^{-1}.

The high rate constant obtained with M = I_2 implies a considerable stability of the I_3 molecule which is in accordance with

9. G. Porter, Z. G. Szabo and M. G. Townsend, *Proc. Roy. Soc.* A270, 493 (1962).

theoretical predictions.[10] In bromine atom recombination reactions the bromine molecule shows a similar efficiency as a chaperon.

Flash-photolysis techniques have also been used to study the iodine atom recombination reaction in non-polar solvents such as carbon tetrachloride.[11] Since the iodine atoms are always adjacent to solvent molecules which can carry away excess kinetic energy, there is no need for a specific chaperon molecule, and the reaction shows straightforward second-order behaviour. In carbon tetrachloride, the activation energy is 3·2 kcal $mole^{-1}$.

Different behaviour is observed in benzene solutions and there is spectral evidence for the formation of a transient intermediate believed to be C_6H_6, I.[12] The disappearance of the latter obeys second-order kinetics and is consistent with the reaction

$$C_6H_6I + C_6H_6I \rightarrow 2C_6H_6 + I_2.$$

The iodine molecule also forms weak charge-transfer complexes with benzene.

The recombination of bromine atoms

Flash-photolysis experiments on the recombination of bromine atoms,[13] although more limited, have produced similar results. Thus third-order rate constants show a similar dependence on the nature of the chaperon and give rise to negative activation energies (Table 7). As before, rate constants k are defined by the rate equation

$$\frac{d[Br_2]}{dt} = k[Br]^2[M].$$

In preliminary experiments on the decay of OH radicals produced by the u.v. flash photolysis of water vapour,[14]

$$OH + H + M \rightarrow$$
$$OH + OH + M \rightarrow,$$

10. G. K. Rollefson and H. Eyring, *J. Am. Chem. Soc.* **54**, 170 (1932).
11. R. L. Strong and J. E. Willard, *J. Am. Chem. Soc.* **79**, 2098 (1957).
12. S. J. Rand and R. L. Strong, *J. Am. Chem. Soc.* **82**, 5 (1960).
13. W. G. Givens and J. E. Willard, *J. Am. Chem. Soc.* **81**, 4773 (1959).
14. G. Black and G. Porter, *Proc. Roy. Soc.* A**266**, 185 (1962).

a similar marked dependence on the chaperon is indicated. Since the OH radical has a comparable electron affinity to that of the halogen atoms, radical–molecule complexes are probably involved.

TABLE 7

KINETIC DATA FOR THE BROMINE ATOM RECOMBINATION WITH DIFFERENT CHAPERONS AT 27°C

M	k (l^2 mole^{-2} sec^{-1}) $\times 10^{-9}$	E (kcal mole^{-1})
He	1·3	
Ar	2·0	−1·4
CO_2	7·8	
Br_2	260	−2·9

The recombination of hydrogen atoms

Flow techniques have been used to study the recombination of hydrogen atoms produced by an electric discharge method. Such experiments are often complicated by an excessive amount of diffusion of hydrogen atoms (rapid diffusion of light atoms) and their recombination on the walls of the reaction vessel, but this can be largely overcome by using a phosphoric acid coated tube when reaction at the surface becomes negligible. At room temperature, and with hydrogen as chaperon, the third-order rate constant is found to be $8{\cdot}9 \times 10^9$ l^2 mole^{-2} sec^{-1}.

Rate constants have also been obtained at much higher temperatures by measuring the recombination of hydrogen atoms in the burnt gases from hydrogen/oxygen/nitrogen burner flames at atmospheric pressure. Although such methods do not permit a wide variation of chaperon molecules, the approach is of particular interest in that a large variation of temperature and composition of the reactants can be achieved whilst interference from reaction vessel walls is absent. The temperature of the flames can be varied over the range 800–2100°C using nitrogen as a diluent gas.[15]

15. E. M. BULEWICZ and T. M. SUGDEN, *Trans. Faraday Soc.* **54,** 1855 (1958); G. DIXON-LEWIS, M. M. SUTTON and A. WILLIAMS, *Discussions Faraday Soc.* 33, 207 (1962).

Since both hydrogen atoms and hydroxyl radicals are produced there are two possible reactions to consider,

$$H + H + M \xrightarrow{k_1} H_2 + M,$$

$$H + OH + M \xrightarrow{k_2} H_2O + M,$$

where in hydrogen rich flames, M can be either H_2, N_2 or H_2O. The decrease in concentration of atoms and radicals with time, i.e. along the direction of flow of the gases, can be determined by adding traces of sodium salt to the flame and measuring the chemiluminescence resulting from reactions

$$H + H + Na \rightarrow H_2O + Na^*,$$

$$H + OH + Na \rightarrow H_2O + Na^*.$$

For flames at 1370°C and with the hydrogen molecule as third-body $k_1 = 2 \cdot 2 \times 10^{10}$ l^2 $mole^{-2}$ sec^{-1} and $k_2 = 5 \cdot 4 \times 10^{11}$ l^2 $mole^{-2}$ sec^{-1}. By varying the relative amounts of N_2, H_2 and H_2O in the burnt gases, these three molecules would appear to have similar chaperon efficiences for reaction k_1. Hydrogen may be twice as efficient a third-body as nitrogen, but bearing in mind the limitations of the experiments, no conclusions have yet been possible regarding the reaction sequence. Formation of charge-transfer complexes is thought to be less likely for atoms of low atomic number which do not readily expand their outer electron shell beyond that of the corresponding inert gas (i.e. there are no *d*-orbitals).

Other atom-recombination reactions

Flow and static systems have been used to measure further three-body recombination rate constants at 27°C,

$O + O + O_2 \rightarrow O_2 + O_2$, ($< 2 \cdot 5 \times 10^8$ l^2 $mole^{-2}$ sec^{-1})

$N + N + N_2 \rightarrow N_2 + N_2$, ($8 \cdot 0 \times 10^8$ l^2 $mole^{-2}$ sec^{-1})

$N + O + N_2 \rightarrow NO + N_2$. ($3 \cdot 3 \times 10^9$ l^2 $mole^{-2}$ sec^{-1})

It is informative to compare these with the corresponding value for the hydrogen atom recombination reaction

$$H + H + H_2 \rightarrow H_2 + H_2. \quad (8{\cdot}9 \times 10^9 \ l^2 \ mole^{-2} \ sec^{-1})$$

since for reactions between like atoms at least, a definite trend in rates is observed.

A statistical mechanical method has been used to calculate such rate constants.[16] Assuming that reactions occur in two steps,

$$X + X \rightleftharpoons X_2^*$$

$$X_2^* + X_2 \rightarrow X_2 + X_2,$$

values obtained agree well with the above experimental results. The large variation in rate constants (a factor of forty in going from hydrogen to oxygen) is attributed to differences in atomic masses and statistical weight ratios rather than to differences in the interatomic potential curves of the various molecules. Re-dissociation of molecules in high vibrational levels would seem to be important in reducing the net rate of recombination.

16. C. B. Kretschmer and H. C. Petersen, *J. Chem. Phys.* **39**, 1772 (1963).

CHAPTER 6

BIMOLECULAR GAS-PHASE REACTIONS

FEW, if any, gas-phase reactions between stable reactants, i.e. substances which can be isolated, are single-stage bimolecular processes. Two reactions which, until recently, were thought to fall into this category,

$$NO_2 + NO_2 \rightarrow 2NO + O_2$$

and

$$H_2 + I_2 \rightarrow 2HI,$$

are now known to react at least in part by alternative routes involving NO_3 radicals and hydrogen and iodine atoms. In the latter the atom path is probably marked even at low temperatures and at 530°C it accounts for $>95\%$ of the hydrogen iodide produced.

All the reactions considered below are, at least in part, by radical or free atom paths. When a radical formed in the first step is continually regenerated with the formation of product, for example, the sequence

$$H_2 + Br \rightarrow HBr + H,$$

$$H + Br_2 \rightarrow HBr + Br,$$

in the reaction between hydrogen and bromine, that reaction is said to have a chain mechanism. The average number of molecules of product formed in a single chain before termination is known as the chain length.

From collision theory, frequency factors for bimolecular reactions are in the region 10^{11}–10^{12} l mole^{-1} sec^{-1}. Experimental values of the pre-exponential A-factor are found to lie within the range 10^6 (for more complex organic molecules) to 10^{12} l mole^{-1} sec^{-1}.

The reaction of hydrogen atoms with molecular hydrogen

The two nuclear spins in the hydrogen molecule can be either parallel (ortho-hydrogen) or anti-parallel (para-hydrogen). Both forms are stable in that there is no direct change,

$$o\text{-H}_2 \rightleftharpoons p\text{-H}_2,$$

but conversion of one to the other takes place readily in the presence of charcoal, paramagnetic substances such as O_2, or with hydrogen atoms. Above 0°C the ratio of ortho- to para-hydrogen in an equilibrium mixture (i.e. in the presence of charcoal) is 3 : 1, which is the highest ortho content possible. At lower temperatures the amount of ortho decreases, and a sample of pure para-hydrogen can be prepared by passing normal hydrogen over charcoal at the boiling point of liquid hydrogen. Thermal conductivity measurements can be used to determine the composition of an unknown mixture.

The conversion of para-hydrogen to a 3 : 1 equilibrium mixture using hydrogen atoms,

$$p\text{-H}_2 + \text{H} \rightleftharpoons o\text{-H}_2 + \text{H},$$

provides a useful means of determining the rate of the reaction between hydrogen atoms and molecular hydrogen. The reaction has been studied by flow methods at room temperature and at pressures between 0·3 and 0·6 mm of Hg. Hydrogen atoms are produced by passing an electric discharge through hydrogen at around 0·5 mm pressure, the latter stream being mixed with para-enriched hydrogen just prior to entering the reaction vessel (volume ~ 1 litre). By measuring the concentration of hydrogen atoms at the entrance and exit to the reactor using Wrede–Harteck gauges,[(1)] the recombination of hydrogen atoms during

1. E. Wrede, *Z. Physik.* **54**, 53 (1929); K. H. Geib and P. Harteck, *Z. Physik Chem., Bodenstein.* 849 (1931).

the 0·4 to 0·9 sec spent in the reactor can be allowed for (the concentration of hydrogen atoms is unaffected by the reaction with molecular hydrogen). After the gases have passed through the reactor, the hydrogen atoms are rapidly converted to hydrogen, by passing through a trap containing platinum gauze at the temperature of liquid air, and the ortho to para content is then

TABLE 8

THE RESISTANCE OF A PLATINUM WIRE CARRYING A FIXED HEATING CURRENT VARIES WITH THE THERMAL CONDUCTIVITY OF THE SURROUNDING HYDROGEN GAS. CALIBRATION OF APPARATUS

Sample of hydrogen	Resistance of wire (ohm)
para-hydrogen	105·30
normal hydrogen	102·72
ortho-deuterium	113·04
normal deuterium	111·47

determined. To do this, the resistance of a platinum wire contained in a vessel of the hydrogen of unknown composition and carrying a fixed heating current is measured. Since the two forms of hydrogen have different thermal conductivities under such conditions, the temperature and therefore the resistance of the platinum wire varies as the composition of the gas is varied. The apparatus is first calibrated using samples of hydrogen of known composition (see Table 8).

The para to ortho conversion can also be studied at around 700°C by non-flow methods, when hydrogen atoms are produced by thermal dissociation.[2] Since the concentration of hydrogen atoms in the reaction sequence

$$H_2 + M \underset{k_{-1}}{\overset{k_1}{\rightleftharpoons}} H + H + M \qquad (1)$$

$$p\text{-}H_2 + H \underset{k_{-2}}{\overset{k_2}{\rightleftharpoons}} o\text{-}H_2 + H, \qquad (2)$$

2. A. FARKAS and L. FARKAS, *Proc. Roy. Soc.* A152, 124 (1935).

is determined by (1) alone, it follows that

$$[H] = (K_1[H_2])^{1/2}$$

The rate of conversion of para-hydrogen can therefore be expressed

$$\frac{-d[p\text{-}H_2]}{dt} = (k_2[p\text{-}H_2] - k_{-2}[o\text{-}H_2])(K_1[H_2])^{1/2},$$

and this can be integrated after substituting for the concentration of ortho-hydrogen,

$$[o\text{-}H_2] = [H_2] - [p\text{-}H_2].$$

A further substitution, $k_{-2} = \frac{1}{3}k_2$, follows from a consideration of the final equilibrium mixture. Note, also, that measured rate constants are only half the actual rate constants, since only half the reactions result in an interconversion. Thus, if the orientation of a nuclear spin is denoted by an arrow, possible reactions of para-hydrogen are

$$\uparrow + \uparrow\downarrow = \uparrow\uparrow + \downarrow \tag{3}$$

$$\uparrow\downarrow + \uparrow = \uparrow + \downarrow\uparrow \tag{4}$$

where in (3) only is there conversion.

Deuterium also has a nuclear spin, but, in this case, the ortho form is obtained at lower temperatures, and at room temperature the equilibrium mixture consists of a 2 : 1 ratio of ortho to para forms. The interconversion of these and the deuterium–hydrogen exchange, overall reaction

$$H_2 + D_2 \rightarrow 2HD,$$

can be followed using similar procedures. A complete list of rate constants is shown in Table 9, where these are about half the values originally reported. In determining the more recent values,[3] extra precautions were necessary to exclude all traces of oxygen, the latter having a marked catalytic effect which is

---

3. G. Boato, G. Careri, A. Cimino, E. Molinari and G. G. Volpi, *J. Chem. Phys.* **24**, 783 (1956).

probably due to the formation of OH radicals. Activation energies lie within the 6·0 ± 1·0 kcal mole^{-1} range. Because of their simplicity the theoretical treatment of these reactions is of particular interest.

From rotational spectra, other diatomic molecules are known to have ortho and para forms, e.g. $^{14}N_2$ and $^{35}Cl_2$. Because the latter have relatively high moments of inertia, limiting high

TABLE 9

RATE CONSTANTS FOR A VARIETY OF REACTIONS BETWEEN HYDROGEN ATOMS AND MOLECULES AT 727°C (IN EACH CASE, ALL POSSIBLE REACTION PATHS HAVE BEEN TAKEN INTO ACCOUNT)

Reaction	$k \times 10^{-9}$ (l mole^{-1} sec^{-1})
$H + H_2 \rightarrow H_2 + H$	1·1
$D + D_2 \rightarrow D_2 + D$	0·57
$D + H_2 \rightarrow HD + H$	0·98
$H + D_2 \rightarrow HD + D$	0·61
$H + HD \rightarrow H_2 + D$	0·37
$D + DH \rightarrow D_2 + H$	0·40

temperature ratios of ortho to para forms persist to temperatures well below the normal boiling points; and enrichments such as have been achieved with hydrogen and deuterium are not therefore possible.

The reaction between hydrogen and iodine

The reaction of hydrogen with iodine, and the reverse, were first studied in the 1890's, over a temperature range 280–500°C. Until recently, both processes were believed to be exclusively by a single-stage bimolecular step involving a transition complex,

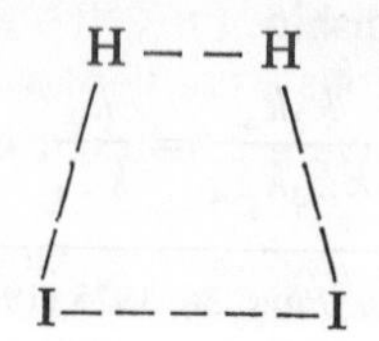

It has now been shown[4] that reaction also proceeds by an atom mechanism, which is at least 10% effective at 360°C and 27% effective at 460°C. The latter involves steps similar to those in the reaction between hydrogen and bromine, the complete mechanism being

$$H_2 + I_2 \underset{k_{-1}}{\overset{k_1}{\rightleftharpoons}} 2HI$$

$$I_2 + M \underset{k_{-2}}{\overset{k_2}{\rightleftharpoons}} 2I + M$$

$$I + H_2 \underset{k_{-3}}{\overset{k_3}{\rightleftharpoons}} HI + H$$

$$H + I_2 \underset{k_{-4}}{\overset{k_4}{\rightleftharpoons}} HI + I.$$

Making the stationary-state approximations

$$\frac{d[I]}{dt} = 0, \text{ and } \frac{d[H]}{dt} = 0,$$

expressions for [H] and [I] can be obtained, thus

$$[I] = \sqrt{K_2[I_2]},$$

and

$$[H] = \sqrt{K_2[I_2]}\,\frac{k_3[H_2] + k_{-4}[HI]}{k_{-3}[HI] + k_4[I_2]}.$$

By substituting these in the rate equation

$$\frac{d[HI]}{dt} = 2k_1[H_2][I_2] - 2k_{-1}[HI]^2 + k_3[H_2][I] - k_{-3}[HI][H] + k_4[H][I_2] - k_{-4}[HI][I],$$

together with the relationship

$$\frac{k_3k_4}{k_{-3}k_{-4}} = \frac{k_4}{k_{-1}},$$

4. J. H. Sullivan, *J. Chem. Phys.* **36**, 1925 (1962), and *J. Chem. Phys.* **46**, 73 (1967).

which is obtained by considering the final equilibrium state, an expression

$$\frac{d[HI]}{dt} = 2(k_1[H_2][I_2] - k_{-1}[HI]^2)\left(1 + \frac{k_3k_4K_2[I_2]}{k_1k_{-3}[HI] + k_1k_4[I_2]}\right),$$

is obtained. Earlier experiments failed, therefore, to detect the fractional term in the second bracket. From an extensive but rather complex treatment of the experimental data, individual rate constants and Arrhenius parameters can be obtained. These are shown in Table 10.

TABLE 10

ARRHENIUS PARAMETERS FOR THE DIFFERENT PATHS IN THE REACTION BETWEEN HYDROGEN AND IODINE AT ~400°C

Reaction	$\log_{10}A$	E (kcal mole^{-1})
$H_2 + I_2 \xrightarrow{k_1}$	11·10	40·7
$HI + HI \xrightarrow{k_{-1}}$	11·42	45·9
$H_2 + I \xrightarrow{k_3}$	11·17	32·8
$H + HI \xrightarrow{k_3}$	—	0
$H + I_2 \xrightarrow{k_4}$	—	0

Direct evidence for hydrogen atom participation can be obtained by following the conversion of para-hydrogen to ortho-hydrogen in an equilibrium mixture of hydrogen, iodine and hydrogen iodide. The observed rate is about twice as fast as would be expected were the direct reaction to hydrogen iodide and the reverse,

$$H_2 + I_2 \rightleftharpoons 2HI,$$

the only reactions. More recent work by Sullivan (1967)[4] has indicated that the bimolecular path does not occur.

The reaction between hydrogen and bromine

The thermal reaction of hydrogen with bromine has been studied over a wide range of temperatures, being most convenient to follow between 230° and 300°C. The empirical rate equation[5] is of the form

$$\frac{d[\mathrm{HBr}]}{dt} = \frac{a[\mathrm{H_2}][\mathrm{Br_2}]^{\frac{1}{2}}}{1 + b[\mathrm{HBr}]/[\mathrm{Br_2}]},$$

where a and b are constants. There is no evidence for the direct bimolecular reaction

$$H_2 + Br_2 \rightarrow 2HBr,$$

since there is no term in $[H_2][Br_2]$. The reverse reaction, the decomposition of HBr, is negligible at these temperatures.

It is unlikely that an empirical rate equation of the above type would be obtained without certain preconceived ideas as to the mechanism. If such were in fact the case, however, the mechanism might be deduced in the following manner. First of all the $[Br_2]^{\frac{1}{2}}$ term indicates bromine atom participation. Thermal dissociation of bromine and not hydrogen as a first step is reasonable, since their bond energies are 46 and 104 kcal mole^{-1}, respectively. Apart from recombination, the only possible reaction of bromine atoms is with hydrogen,

$$H_2 + Br \rightarrow HBr + H,$$

when hydrogen atoms are produced. Terms on the lower line of rate equations generally represent competing reactions, one of them often being a back reaction. In this particular case, there are, in effect, two terms, in hydrogen bromide and bromine, the only species which these can compete for being hydrogen atoms; hence the full mechanism

$$Br_2 + M \underset{k_{-1}}{\overset{k_1}{\rightleftharpoons}} 2Br + M$$

$$Br + H_2 \underset{k_{-2}}{\overset{k_2}{\rightleftharpoons}} HBr + H$$

$$H + Br_2 \xrightarrow{k_3} HBr + Br.$$

5. M. Bodenstein and S. C. Lind, *Z. Physik Chem.* **57**, 168 (1906).

Assuming stationary-state kinetics for [H] and [Br],

$$[H] = \frac{k_2[Br][H_2]}{k_3[Br_2] + k_{-2}[HBr]},$$

and

$$[Br] = (K_1[Br_2])^{1/2}.$$

Substituting both these into the rate expression for $d[HBr]/dt$ (see pages 49–50 for details), a rate equation

$$\frac{d[HBr]}{dt} = \frac{2K_1{}^{1/2}k_2[H_2][Br_2]^{1/2}}{1 + k_{-2}[HBr]/k_3[Br_2]}$$

is obtained,[6] which is of the same form as the experimental equation. The constants a and b can therefore be identified as

$$a = 2k_2K_1{}^{1/2},$$

and

$$b = k_{-2}/k_3.$$

The constant $b = 0{\cdot}12$ is small and does not vary appreciably with temperature, so that by having the concentration of bromine at all times in excess of that of hydrogen bromide, the rate equation reduces to the form

$$d[HBr] = 2K_1{}^{1/2}k_2[H_2][Br_2]^{1/2}.$$

Under these conditions $K_1{}^{1/2}k_2$ can be determined, and since the equilibrium constant K_1 for the dissociation of bromine is known from spectroscopic data, k_2 and, from the temperature dependence, E_2 can be obtained ($E_2 = 17{\cdot}6$ kcal mole^{-1}). Knowing E_2 and using the thermal data,

$$H_2 + Br \underset{k_{-2}}{\overset{k_2}{\rightleftharpoons}} HBr + H - 16{\cdot}6 \text{ kcal mole}^{-1},$$

it follows that $E_2 = 1{\cdot}0$ kcal mole^{-1}. Furthermore, since the experimental quantity $b = k_{-2}/k_3$ varies little with temperature (~5% over the range 25–300°C) E_3 must be of the same order

6. See, e.g., M. Polanyi, *Z. Electrochem.* **26**, 50 (1920); J. A. Christiansen, *Kgl. Danske Vidensk. Selsk. Mat.-fys., Medd.* **1**, 14 (1919); and K. F. Herzfeld, *Z. Elektrochem.* **25**, 30 (1919).

as E_{-2}, and E_{-3} can therefore be inferred even though k_{-3} is not effective in this system (see Table 11). The general picture that emerges is, therefore, one of a relatively slow chain propagation step,

$$H_2 + Br \xrightarrow{k_2} HBr + H, \ (E_2 = 17{\cdot}6),$$

followed by the rapid reaction of hydrogen atoms, i.e. E_{-2} and E_3 are small. The back reaction k_{-3} is not effective because of the high activation energy.

TABLE 11

ARRHENIUS PARAMETERS FOR THE DIFFERENT PATHS IN THE REACTION OF HYDROGEN WITH BROMINE AT 300°C

Reaction	log A	E (kcal mole^{-1})
$Br + H_2 \xrightarrow{k_2}$	10·68	17·6
$H + Br_2 \xrightarrow{k_{-2}}$	11·20	1·0
$H + HBr \xrightarrow{k_3}$	10·33	1·0
$HBr + Br \xrightarrow{k_{-3}}$	10·50	41·8

Bromine atoms can also be produced by photochemical means,

$$Br_2 + I_{obs} \xrightarrow{k} 2Br,$$

where I_{abs} is the intensity of the absorbed light. Subsequent reactions are as in the thermal reaction

$$2Br + M \xrightarrow{k_{-1}} Br_2 + M$$

$$Br + H_2 \underset{k_{-2}}{\overset{k_2}{\rightleftharpoons}} HBr + H$$

$$H + Br_2 \xrightarrow{k_3} HBr + Br.$$

Assuming the recombination of bromine atoms at the walls of the reaction vessel to be negligible, the stationary-state treatment gives

$$[Br] = \frac{(kI_{abs})^{1/2}}{k_{-1}[M]},$$

which replaces

$$[Br] = (K_1[Br_2])^{1/2},$$

in the thermal reaction. Otherwise the treatment is as before.

The reaction between hydrogen and chlorine

The reaction between hydrogen and chlorine is extremely sensitive to wall effects and to traces of impurities, e.g. water and oxygen, and neither thermal nor photochemical studies have proved very satisfactory from the kinetic point of view. Certain features of the thermal reaction have been established however, the most important being that the basic chain sequence

$$Cl_2 + M \rightleftharpoons Cl + Cl + M$$
$$Cl + H_2 \rightleftharpoons HCl + H$$
$$H + Cl_2 \rightleftharpoons HCl + Cl,$$

is so rapid that, once under way (temperatures in the region 200–300°C), the chlorine atom recombination reaction is no longer effective in controlling the concentration of chlorine atoms. In other words, the chlorine atom recombination is slow compared to other reaction steps, and the stationary-state assumption $d[Cl]/dt = 0$ is no longer valid. The addition of nitrosyl chloride to the system at 250–300°C produces an induction period, during which the combination of hydrogen and chlorine is negligible. This is caused by the removal of chlorine atoms

$$NOCl + Cl \rightarrow NO + Cl_2,$$

and demonstrates, conclusively, that the bimolecular reaction

$$H_2 + Cl_2 \rightarrow 2HCl$$

is negligible over the range of temperature.

That the overall rate is so much faster than those for the corresponding reactions of hydrogen with iodine and bromine is attributable to the much faster chain propagation step

$$H_2 + Cl \rightarrow HCl + H.$$

The latter has been measured directly with chlorine atoms from a high frequency discharge tube, the activation energy being quite small, i.e. 5·0 kcal mole^{-1}. In the halogen reactions already considered, the activation energies for the corresponding steps

$$H_2 + Br \rightarrow HBr + H$$

and

$$H_2 + I \rightarrow HI + H$$

are 17·6 and 32·8 kcal/mole^{-1} respectively.

Chlorine absorbs light of wavelength less than 550 mμ,

$$Cl_2 + h\nu \rightarrow 2Cl$$

and a photochemical reaction results. Quantum yields are rather variable but, with light of wavelength less than 478 mμ, chain lengths of from 10^4 to 10^5 have been observed in normal reaction vessels at room temperature, i.e. for each light quantum absorbed, from 10^4 to 10^5 molecules of hydrogen chloride are produced. Most kinetic investigations have been with such photochemical systems. At temperatures below −70°C the chain cycle is found to be appreciably slower and under these conditions the chlorine atom recombination reaction becomes effective. At −100°C the kinetics of the system are reproducible and comparatively well behaved.[7]

The reaction between hydrogen and fluorine

Trends observed in the reaction of hydrogen with chlorine are continued in the reaction of hydrogen and fluorine. Thus the reaction with fluorine is not only fast but highly exothermic

---

7. J. C. Potts and G. K. Rollefson, *J. Am. Chem. Soc.* **57**, 1027 (1935).

(Table 12), and equivalent amounts of hydrogen and fluorine are explosive at atmospheric pressure. Under non-explosive conditions, i.e. at reduced pressures, the reaction is surface sensitive and is difficult to interpret. It can be assumed that fluorine atoms initiate the reaction since the bond energy for F_2 is only 36 kcal mole^{-1}. Subsequently, fluorine atoms would seem to be

TABLE 12

THE OVERALL EXOTHERMICITY OF HYDROGEN–HALOGEN REACTIONS

Reaction	ΔH (kcal mole^{-1})
$H_2 + F_2 \rightarrow 2HF$	−128
$H_2 + Cl_2 \rightarrow 2HCl$	−44
$H_2 + Br_2 \rightarrow 2HBr$	−24
$H_2 + I_2 \rightarrow 2HI$	−3

just as reactive as hydrogen atoms, so that chain termination reactions

$$H + H + M \rightarrow H_2 + M$$

and

$$H + F + M \rightarrow HF + M$$

can no longer be ignored.

The decomposition of nitrogen dioxide

The derivation and treatment of the rate equation for the concurrent molecular and free radical paths in the decomposition of nitrogen dioxide has already been considered (pp. 54–57).[8] Over the first 10% of the reaction at temperatures *ca.* 400°C, the radical path

$$NO_2 + NO_2 \underset{k_{-2}}{\overset{k_2}{\rightleftharpoons}} NO_3 + NO$$

$$NO_3 + NO_2 \xrightarrow{k_3} NO + O_2 + NO_2,$$

8. P. G. ASHMORE and M. G. BURNETT, *Trans. Faraday Soc.* **58**, 253 (1962).

is effective and rates of up to twice those expected from the straightforward bimolecular reaction

$$NO_2 + NO_2 \xrightarrow{k_1} 2NO + O_2$$

are observed. As the reaction proceeds, the concentration of nitric oxide increases and the NO_3 path becomes negligible.

Activation energies corresponding to k_1 and k_2 are 26·9 and 23·9 kcal mole^{-1} respectively. Since the NO_3 radical is thought to have the same configuration as the nitrate ion, transition complexes may be formulated as follows

$$\text{O–N–O} + \text{O–N–O} \rightleftharpoons$$
$$\text{O–N}\cdot\cdot\text{O}\cdot\cdot\text{O}\cdot\cdot\text{N–O} \rightarrow \text{O–N} + O_2 + \text{N–O}$$

$$\text{O–N–O} + \text{N}(O)_3 \rightleftharpoons \text{O–N}\cdot\cdot\text{O}\cdot\cdot\text{N}(O)_2 \rightarrow \text{O–N} + NO_3.$$

It is of interest to compare these with the further reaction of two nitrogen dioxide molecules at lower temperatures,

$$O_2N + NO_2 \rightarrow O_2N - NO_2$$

For the latter, the activation energy is small and close to zero.

The decomposition of nitrosyl chloride

The thermal decomposition of nitrosyl chloride, like that of nitrogen dioxide, was originally thought to be a single-stage bimolecular process,

$$2ClNO \rightarrow Cl_2 + 2NO.$$

Evidence contrary to this was obtained by working at a higher temperature, ~300°C, when rates were a good deal faster than those expected.[9] The difference can be accounted for by considering concurrent radical reactions. Thus the dissociation

$$\text{ClNO} + \text{M} \rightarrow \text{NO} + \text{Cl} + \text{M},$$

is followed by the fast reaction

$$\text{ClNO} + \text{Cl} \rightarrow \text{NO} + \text{Cl}_2.$$

The addition of chlorine catalyses the decomposition due to the resultant increase in the number of chlorine atoms.

The reaction between carbon monoxide and chlorine

The rate of formation and decomposition of phosgene, overall equation

$$\text{CO} + \text{Cl}_2 \rightleftharpoons \text{COCl}_2,$$

have been studied separately in the 350–450°C temperature range.[10] The rate law for the initial stages of the forward reaction was found to be

$$\frac{d[\text{COCl}_2]}{dt} = a[\text{Cl}_2]^{3/2}[\text{CO}],$$

where a is a constant. A rate equation of this form can be derived for the mechanism

$$\text{Cl}_2 + \text{M} \underset{k_{-1}}{\overset{k_1}{\rightleftharpoons}} 2\text{Cl} + \text{M}$$

$$\text{Cl} + \text{CO} + \text{M} \underset{k_{-2}}{\overset{k_2}{\rightleftharpoons}} \text{COCl} + \text{M}$$

$$\text{COCl} + \text{Cl}_2 \xrightarrow{k_3} \text{COCl}_2 + \text{Cl},$$

9. P. G. Ashmore and J. Chanmugan, *Trans. Faraday Soc.* **49**, 25 (1953).
10. W. G. Burns and F. S. Dainton, *Trans. Faraday Soc.* **48**, 39 (1952).

by making stationary-state approximations

$$\frac{d[\mathrm{Cl}]}{dt} = 0, \text{ and } \frac{d[\mathrm{COCl}]}{dt} = 0.$$

Thus, by assuming that $k_{-2}[\mathrm{M}] \gg k_3[\mathrm{Cl_2}]$, it can be shown that

$$[\mathrm{Cl}] = K_1{}^{1/2}[\mathrm{Cl_2}]^{1/2},$$

and

$$[\mathrm{COCl}] = K_1{}^{1/2}K_2[\mathrm{Cl_2}]^{1/2}[\mathrm{CO}],$$

hence

$$\frac{d[\mathrm{COCl_2}]}{dt} = k_3K_1{}^{1/2}K_2[\mathrm{Cl_2}]^{3/2}[\mathrm{CO}].$$

Which is in agreement with the experimental rate equation.

For the initial stages of the decomposition of phosgene in the presence of chlorine, the rate law is of the form

$$\frac{-d[\mathrm{COCl_2}]}{dt} = b[\mathrm{Cl_2}]^{1/2}[\mathrm{COCl_2}].$$

By considering the back reaction k_{-3} instead of k_3, i.e. a mechanism

$$\mathrm{Cl_2} + \mathrm{M} \underset{k_{-1}}{\overset{k_1}{\rightleftharpoons}} 2\mathrm{Cl} + \mathrm{M}$$

$$\mathrm{Cl} + \mathrm{CO} + \mathrm{M} \underset{k_{-2}}{\overset{k_2}{\rightleftharpoons}} \mathrm{COCl} + \mathrm{M}$$

$$\mathrm{COCl_2} + \mathrm{Cl} \xrightarrow{k_{-3}} \mathrm{COCl} + \mathrm{Cl_2},$$

and again assuming stationary-state kinetics for Cl and COCl, the initial rate is

$$\frac{-d[\mathrm{COCl_2}]}{dt} = k_{-3}K_1{}^{1/2}[\mathrm{Cl_2}]^{1/2}[\mathrm{COCl_2}].$$

Note that for the intermediate stages the rate law is

$$\frac{d[\mathrm{COCl_2}]}{dt} = k_3K_1{}^{1/2}K_2[\mathrm{Cl_2}]^{3/2}[\mathrm{CO}] - k_{-3}K_1{}^{1/2}[\mathrm{Cl_2}]^{1/2}[\mathrm{COCl_2}],$$

and that at equilibrium,

$$k_3K_1^{1/2}K_2[Cl_2]^{3/2}[CO] = k_{-3}K_1^{1/2}[Cl_2]^{1/2}[COCl_2],$$

i.e.

$$\frac{[COCl_2]}{[CO][Cl_2]} = K,$$

where K is the overall equilibrium constant.

An alternative mechanism involving intermediate Cl_3 has been considered,

$$Cl_2 + M \rightleftharpoons Cl + Cl + M$$

$$Cl_2 + Cl + M \rightleftharpoons Cl_3 + M$$

$$Cl_3 + CO \rightleftharpoons COCl_2 + Cl,$$

since it also gives rate equations of the required form. From a kinetic study of the thermal reaction alone, a decision cannot be made between the two. Investigations of the photo-chemical reaction between carbon monoxide and chlorine suggest that COCl is a more likely intermediate than Cl_3.

The reaction between hydrogen and oxygen

Although the reaction between hydrogen and oxygen is difficult to interpret quantitatively, the mechanism is now fairly well understood.[11] One of the main difficulties has been the surface sensitive nature of the reaction, and the recent history of the surface of a reaction vessel as well as its surface area to volume ratio (a spherical vessel has least surface area per unit volume) will often affect the rate. The first of these is partly overcome by using Pyrex reaction vessels coated with a salt such as potassium chloride, but, even so, reproducible results are not always obtained, and the shape, volume and details of the surface of the reactor should always be given.

At lower temperatures, the reaction is slow unless a surface catalyst such as platinum is present. For example, below 400°C,

---

11. See, e.g., G. VAN ELBE and B. LEWIS, *J. Chem. Phys.* **10**, 366 (1942), and C. N. HINSHELWOOD and A. T. WILLIAMSON, *The Reaction between Hydrogen and Oxygen*, Oxford University Press, 1934.

reaction of a 2 : 1 hydrogen to oxygen mixture in a spherical reaction vessel, volume ~220 ml, is very slow. At only slightly higher temperatures, however, the system becomes explosive, due to the exothermic nature of the overall reaction

$$2H_2 + O_2 \rightarrow 2H_2O + 115{\cdot}6 \text{ kcal mole}^{-1}.$$

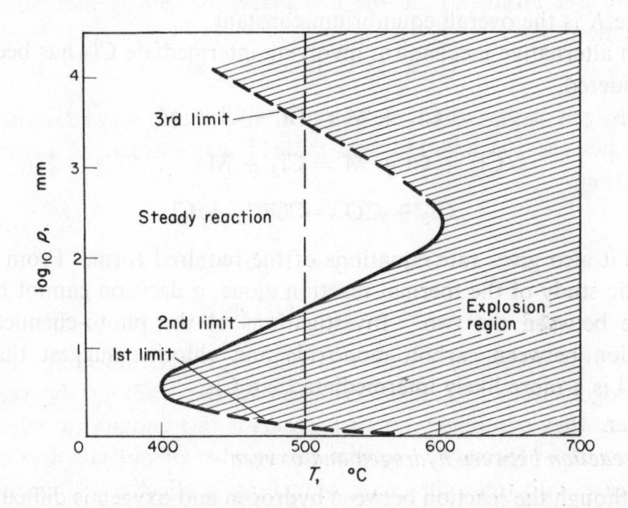

FIG. 23. Explosion limits for a 2 : 1 hydrogen to oxygen mixture in a spherical Pyrex (KCl-coated) reaction vessel, volume ~220 ml. The logarithm of the pressure p is used for convenience.

Above 600°C it is explosive at all pressures, but between 400° and 600°C the explosion is effectively quenched over a certain pressure range, as shown in Fig. 23. Transitions from the region of slow reaction to that of explosion are quite abrupt. Consider, for instance, a 2 : 1 hydrogen to oxygen mixture at 500°C (broken line in Fig. 23). If the pressure is below the first explosion limit of 2 mm of Hg ($\log_{10}p = 1{\cdot}3$), reaction is very slow indeed, and probably confined to the walls of the vessel. At higher pressures up to 25 mm ($\log_{10}p = 1{\cdot}4$), the mixture is explosive, then, between the second and third explosion limits the rate is slow, and finally, at pressures above 3160 mm ($\log_{10}p = 3{\cdot}5$), the mixture is once again explosive.

Explosions are the result of chain branching reactions,

$$H + O_2 \rightarrow OH + O,$$

where free atoms are, in the first place, formed by thermal dissociation,

$$H_2 + M \rightarrow H + H + M.$$

Hydrogen rather than oxygen will dissociate, since the bond energies are 104 and 118 kcal mole^{-1} respectively. A cycle of reactions

$$H + O_2 \rightarrow OH + O$$

$$OH + H_2 \rightarrow H_2O + H$$

$$O + H_2 \rightarrow OH + H,$$

gives two hydrogen atoms and one hydroxyl radical for each hydrogen atom. If, then, diffusion to the walls of the vessel (where they combine with each other) is fast enough to balance their rate of formation, so that the number of radicals does not increase with time, the chain branching is effectively quenched. The reaction is controlled in this way for pressures up to the first explosion limit, but, at higher pressures, chain branching is effective, the reaction becoming immeasurably fast, i.e. explosive (there is a bright flash and a sharp sound). This explanation is supported by experiment, since (a), in a larger reaction vessel the first explosion limit is lowered and (b), with the addition of inert gases the first explosion limit is also lowered. Thus it can be concluded that termination of the chain branching process is diffusion controlled.

The second pressure limit is dependent on the total pressure of the system but not the size of the reaction vessel. Since the addition of an inert gas also helps to quench the explosion, the reaction

$$H + O_2 + M \rightarrow HO_2 + M$$

seems likely, and under these conditions is thought to replace the chain branching reaction

$$H + O_2 \rightarrow HO + O.$$

The HO_2 radicals so formed can either diffuse to the walls where they react to form stable species, or react with hydrogen,

$$HO_2 + H_2 \rightarrow H + H_2O_2,$$

or

$$HO_2 + H_2 \rightarrow OH + H_2O.$$

Diffusion to the walls is believed to be dominant over the region of slow reaction between the second and third explosion limits, but, as the pressure is increased, further diffusion is so much impeded that reaction with hydrogen becomes more probable. Chain branching again becomes effective and the third explosion limit is reached.

CHAPTER 7

ISOTOPIC EXCHANGE REACTIONS IN AQUEOUS SOLUTIONS

ELECTRON-TRANSFER reactions of the type

$$Fe^{2+} + {}^*Fe^{3+} \rightarrow Fe^{3+} + {}^*Fe^{2+} \quad (1)$$

$$^*Co(phen)_3^{2+} + Co(phen)_3^{3+} \rightarrow {}^*Co(phen)_3^{3+} + Co(phen)_3^{2+} \quad (2)$$

$$^*MnO_4^{2-} + MnO_4^- \rightarrow {}^*MnO_4^- + MnO_4^{2-} \quad (3)$$

are generally studied by isotopic labelling of one or other of the reactants as indicated. For the faster reactions, NMR and ESR techniques can sometimes be used. Such reactions constitute a relatively simple class of reaction in that

(*a*) there is no net chemical change,

(*b*) rate constants for forward and back reactions are to all intents equal, so that the equilibrium constant K may be assumed unity and

(*c*) the standard free energy change ΔG_0 may be assumed to be zero.†

For electrons to be transferred, however, activation energy requirements must be met as in all other reactions. In particular, energy is required to reorganize the coordination and solvation spheres around each of the reactants, otherwise, in (1) above, for example, electron transfer would result in the formation of a

† The even distribution of an isotope between two oxidation states is accompanied by a small and positive entropy change, so that, in actual fact, there is a *small* free energy change.

ferric ion in a ferrous hydration sphere (Fe—O distance 2·2 Å) and a ferrous ion in a ferric hydration sphere (Fe—O distance 2·05 Å)

$$Fe^{2+} + Fe^{3+} \longrightarrow Fe^{3+} + Fe^{2+}$$

Experimental procedures for (1) and (3), which may be considered typical, are outlined in Chapter 2. For reactions between cations, perchlorate salts are used whenever possible since, of the simple anions, perchlorate shows least tendency to complex. Unless otherwise stated it should be assumed that all reactions are in a perchlorate medium. Some closely related reactions of the type

$$V^{III} + V^{V} \rightarrow 2V^{IV},$$

$$Fe^{2+} + Fe(bipy)_3^{3+} \rightarrow Fe^{3+} + Fe(bipy)_3^{2+},$$

in which there is a net chemical change and ΔG_0 is negative are also considered in this chapter.

The closeness of approach of the reactants in the majority of electron transfer reactions is uncertain. In a few cases, such as the Cr^{2+}–$Cr^{III}X^{3+}$ reactions where $X^- = F^-, Cl^-, Br^-$, etc., *inner-sphere* transition complexes of the type

$$[(H_2O)_5Cr^{II}\,X\,Cr^{III}(H_2O)_5]^{4+}$$

are formed, since one of the products is $Cr(H_2O)_5X^{2+}$ and not $Cr(H_2O)_6^{3+}$ and the reaction

$$Cr(H_2O)_6^{3+} + X^- \rightarrow Cr(H_2O)_5X^{2+} + H_2O,$$

is known to be slow at room temperature. When both reactants are non-labile, however, as in the $Fe(CN)_6^{4-}$–$Fe(CN)_4^{3-}$ exchange, a close approach of the two metal atoms is not possible, and they must always be separated by at least two ligand groups. Such reactions are said to proceed by an *outer-sphere* mechanism.

With non-transition elements, e.g. Tl, Sn, Sb and As, simultaneous two-electron transfers are possible, since stable oxidation states differ by two electrons. Thus, direct one-stage processes

$$Tl^{I} + {}^{*}Tl^{III} \rightarrow Tl^{III} + {}^{*}Tl^{I}$$

$$Sn^{II} + Tl^{III} \rightarrow Sn^{IV} + Tl^{I},$$

in which two electrons from a single orbital on one of the reactants are transferred to a single orbital on the other are possible. From the kinetic point of view, however, the transfer of two electrons one after the other in one encounter is difficult to distinguish from a simultaneous transfer. For the reaction

$$U^{IV} + Tl^{III} \rightarrow U^{VI} + Tl^{I},$$

simultaneous transfer of two electrons would seem less likely, though not impossible, since the two electrons which are transferred from the U^{IV} are in different orbitals and spin pairing must occur at some stage.

A. TRANSITION-METAL IONS

Exchange reactions between different oxidation states of vanadium

In acid perchlorate solutions, the vanadium-(II),-(III),-(IV) and -(V) oxidation states are present as V^{2+}, V^{3+}, VO^{2+} and VO_2^+ ions, respectively. Further hydrolysis of these is not extensive, e.g.

$$V^{3+} + H_2O \rightleftharpoons VOH^{2+} + H^{+} \; (K \sim 2 \times 10^{-3}),$$

but since partially hydrolysed ions such as VOH^{2+} are generally more reactive than the parent cation, small concentrations are often kinetically significant.

This is, in fact, true of the V^{II}–V^{III} exchange.[1] The rate may be expressed $k_{obs}[V^{II}][V^{III}]$, where k_{obs} shows an inverse dependence on the hydrogen-ion concentration,

$$k_{obs} = a + b[H^+]^{-1}.$$

This would suggest that there are two reaction paths,

$$V^{2+} + V^{3+} \xrightarrow{k_1} \qquad (1)$$

$$V^{2+} + VOH^{2+} \xrightarrow{k_2} \qquad (2)$$

The $V^{2+} + VOH^{2+}$ path is assumed relevant rather than the $VOH^+ + V^{3+}$ path since V^{3+} is more extensively hydrolysed than V^{2+}. The product of the concentrations $[V^{2+}][VOH^{2+}]$, is, therefore, much larger than the term $[VOH^+][V^{3+}]$, and, as long as the rate constants for these two paths are not widely different, the $V^{2+} + VOH^{2+}$ path will predominate. In all such instances where the hydrolysis of the two principal reactants is markedly different (hydrolysis constants differ by, say, 10^2), similar assumptions are generally made. From eqns. (1) and (2) above, k_{obs} can be expressed

$$k_{obs} = k_1 + k_2K[H^+]^{-1},$$

where K is for the hydrolysis of V^{3+},

$$V^{3+} + H_2O \rightleftharpoons VOH^{2+} + H^+,$$

and the experimental quantities a and b can be identified as k_1 and k_2K respectively. If the equilibrium constant K is known from separate stability constant measurements, both the true rate constants k_1 and k_2 can be evaluated. Kinetic data for both these paths and for other vanadium reactions are shown in Table 13.

The V^{III}–V^{IV} exchange has just one path showing an inverse acid dependence. This is assumed to be

$$VOH^{2+} + VO^{2+} \xrightarrow{k_3}$$

---

1. K. V. Krishnamurty and A. C. Wahl, *J. Am. Chem. Soc.* **80**, 5921 (1958).

since $V^{3+}(K = 2 \times 10^{-3})$ is more hydrolysed than $VO^{2+}(K = 4{\cdot}4 \times 10^{-6})$. Prior hydrolysis of V^{3+} will obviously favour its conversion to VO^{2+}, but even so, it is, perhaps, surprising that this path should be so very much more favourable than the $V^{3+} + VO^{2+}$ acid independent path.

TABLE 13

KINETIC DATA FOR VANADIUM EXCHANGE REACTIONS AT 25°C

Oxidation states	Actual reactants	k (l mole^{-1} sec^{-1})	$\Delta S\ddagger$ (e.u.)	$\Delta H\ddagger$ (kcal mole^{-1})
$V^{II} + V^{III}$	$V^{2+} + V^{3+} \xrightarrow{k_1}$	$1{\cdot}0 \times 10^{-2}$	−25	12·6
	$V^{2+} + VOH^{2+} \xrightarrow{k_2}$	~1·8	—	—
$V^{III} + V^{IV}$	$VOH^{2+} + VO^{2+} \xrightarrow{k_3}$	1·0	−24	10·7
$V^{IV} + V^{V}$	see text	v. fast	—	—
$V^{III} + V^{V}$	$V^{3+} + VO_2{}^{+} \xrightarrow{k_5}$ (and 3 other paths)	30·1	24·3	8·3
$V^{II} + V^{IV}$	$V^{2+} + VO^{2+} \xrightarrow{k_9}$ (and 1 other path)	1·58	−16·5	12·3

The V^{IV}–V^{V} reaction is too fast to be followed using conventional methods.[2] Nuclear magnetic resonance line broadening techniques have indicated a rate equation of the form $k[V^{IV}][V^{V}]^2$, and an exchange between V^{IV} and a dimeric form of V^{V} seems likely. This is not unreasonable, since V^{V} is known to form a variety of polymeric species, e.g.

$$2VO_2^+ + 2H^+ \rightarrow V_2O_3^{4+} + H_2O.$$

The moderately fast oxidation–reduction reaction between V^{III} and V^{V},

$$V^{III} + V^{V} \rightarrow 2V^{IV},$$

2. C. R. GUILIANO and H. M. MCCONNELL, *J. Inorg. Nucl. Chem.* **9,** 171 (1959).

has been studied spectrophotometrically over the temperature range 0·2° to 34·2°C.[3] Vanadium(V) has a high absorption in the 300–350 mμ region, and by using low reactant concentrations ($\sim 5 \times 10^{-4}$ M), the reaction can be followed with only slight modifications of the normal procedure. The rate law is

$$\frac{d[V^{V}]}{dt} = k'_{obs}[V^{III}][V^{V}],$$

where k'_{obs} shows a rather complex acid dependence,

$$k'_{obs} = k_4[H^+] + k_5 + k_6[H^+]^{-1} + k_7[H^+]^{-2}.$$

In this equation, k_5 corresponds to the reaction of V^{3+} and VO_2^+, while for k_4 the H^+ is probably brought into the transition complex by the VO_2^+, i.e. there is a rapid pre-equilibrium between H^+ and VO_2^+. For paths k_6 and k_7 both the reactants are probably effective in bringing hydroxyl ions into the transition complex. Of the four terms, $k_6[H^+]^{-1}$ predominates over most of the 0·02 to 2·0 M perchloric acid range studied ($\mu = 2·0$ M).

In the reaction of V^{II} with V^{IV},

$$V^{II} + V^{IV} \rightarrow 2V^{III},$$

the acid dependence is of the form

$$k''_{obs} = k_8[H^+] + k_9.$$

Path k_9 is predominant in 0·2 to 2·0 N acid perchlorate solutions ($\mu = 2·0$ M), and at 25°C $k_8 = 0·05$ and $k_9 = 1·58$ l mole^{-1} sec^{-1}.[4] A highly coloured (brown) intermediate VOV^{4+}, which is a hydrolytic dimer of V^{III}, has been identified, and from a study of its rate of formation and disappearance it can be concluded that 65% of the reaction is with the formation of this intermediate. The rest of the reaction proceeds direct to the final products, possibly by way of an outer-sphere activated

3. N. A. Daugherty and T. W. Newton, *J. Phys. Chem.* **68**, 612 (1964).
4. T. W. Newton and F. B. Baker, *J. Phys. Chem.* **68**, 228 (1964).

complex. Such inner- and outer-sphere paths are possible in a number of reactions, although specific intermediates, as in the present case, are not always detectable.

Exchange reactions between different oxidation states of chromium

Hexaquo Cr^{III} exchanges water molecules with solvent only very slowly (half-life *ca.* 40 hr), and substitution of H_2O by X^-, or X^- by H_2O, in the inner coordination sphere of the $Cr^{III}X^{2+}$ complex is similarly slow. Chromium(II), on the other hand, is extremely labile and substitution of its inner coordination sphere is much too fast to be followed by conventional experimental techniques. In acid solutions, the hydrolysis of Cr^{2+} is small so that the concentration of Cr^{2+} may be assumed equal to that of the Cr^{II}. In studying the exchange between Cr^{2+} and CrX^{2+},[5] where the Cr^{2+} is labelled, the Cr^{III} species can be separated by normal precipitation methods with their inner coordination spheres intact. As the exchange proceeds, the amount of labelled chromium separated as Cr^{III} increases but the amount of X^- separated along with the Cr^{III} remains constant. In other words, when electron transfer takes place the X^- group is transferred to the newly formed Cr^{III} ion, i.e. if $AgNO_3$ is added to a reaction solution when $X^- = Cl^-$ there is no immediate precipitation of AgCl. Any other mechanism in which X^- is not transferred would result in a Cr^{2+} induced dissociation of X^- from the original Cr^{III} complex. When the Cr^{2+}–$CrCl^{2+}$ exchange is studied in the presence of free $*Cl^-$, the latter has little, if any, effect on the rate and remains essentially uncoordinated.

For the reactions listed in Table 14, inner-sphere mechanisms have been demonstrated for all but the acid independent Cr^{2+}–Cr^{3+} reaction, which is often assumed to have an outer-sphere mechanism. Its slow rate, $k \leqslant 2 \times 10^{-5}$ l mole^{-1} sec^{-1}, is attributed to the fact that an e_g electron is transferred

$$*Cr^{2+}(t_{2g}^3e_g^1) + Cr^{3+}(t_{2g}^3) \rightarrow *Cr^{3+}(t_{2g}^3) + Cr^{2+}(t_{2g}^3e_g^1).$$

---

5. See, e.g. D. L. Ball and E. L. King, *J. Am. Chem. Soc.* **80**, 1091 (1958).

Contrast the much faster reaction of V^{2+} with V^{3+}($k = 0{\cdot}01$ l mole^{-1} sec^{-1}) and Fe^{2+} with Fe^{3+} ($k = 4{\cdot}20$ l mole^{-1} sec^{-1}), which are also thought to be by outer-sphere mechanisms in which t_{2g} electrons are transferred,

$$V^{2+}(t_{2g}^3) + {}^*V^{3+}(t_{2g}^2) \rightarrow V^{3+}(t_{2g}^2) + {}^*V^{2+}(t_{2g}^3),$$

$$Fe^{2+}(t_{2g}^4e_g^2) + {}^*Fe^{3+}(t_{2g}^3e_g^2) \rightarrow Fe^{3+}(t_{2g}^3e_g^2) + {}^*Fe^{2+}(t_{2g}^4e_g^2).$$

TABLE 14

KINETIC DATA FOR SOME Cr^{II}–Cr^{III} EXCHANGE REACTIONS

Reactants	Temp. (°C)	k (l mole^{-1} sec^{-1})	$\Delta S^\ddagger$ (e.u.)	$\Delta H^\ddagger$ (kcal mole^{-1})
$Cr^{2+} + Cr^{3+}$	25	$\leqslant 2 \times 10^{-5}$	(−8)	(21)
$Cr^{2+} + CrOH^{2+}$	25	0·7	—	—
$Cr^{2+} + CrF^{2+}$	0	$2{\cdot}6 \times 10^{-3}$	−20	13·7
$Cr^{2+} + CrF_2^+$ (*cis*)	0	$1{\cdot}2 \times 10^{-3}$	−24	13·0
$Cr^{2+} + CrCl^{2+}$	0	9	—	—
$Cr^{2+} + CrCl_2^+$ (*trans*)	2	2×10^3	—	—
$Cr^{2+} + CrBr^{2+}$	0	>60	—	—
$Cr^{2+} + CrNCS^{2+}$	27	$1{\cdot}8 \times 10^{-4}$	—	—
$Cr^{2+} + CrN_3^{2+}$	0	1·23	−22·8	9·6
$Cr^{2+} + Cr(N_3)_2^+$ (*cis*)	0	60	—	—

These rates are in accord with ligand field effects since e_g electrons are in orbitals in the direction of the ligand groups, and will have a much greater effect on metal–ligand bond distances than t_{2g} electrons. Activation energy requirements should, therefore, be greater and rates slower for an e_g electron transfer.

In the Cr^{2+}–CrN_3^{2+} and Cr^{2+}–$CrNCS^{2+}$ reactions, where N_3^- and NCS^- are both linear, polyatomic bridged activated complexes are formed. The much slower rate for the $CrNCS^{2+}$ reaction can be attributed to the unsymmetrical nature of the transition complex,

$$[(H_2O)_5\text{—}Cr^{III}\text{—}N\text{—}C\text{—}S\text{—}Cr^{II}(H_2O)_5]^{4+},$$

compared with

$$[(H_2O)_5\text{—}Cr^{III}\text{—}N\text{—}N\text{—}N\text{—}Cr^{II}(H_2O)_5]^{4+}.$$

Electron transfer, in the former, leads to the formation of unstable $CrSCN^{2+}$ in which the thiocyanate group is coordinated through the sulphur atom.

The Cr^{2+}–Cr^{3+} exchange is catalysed by the addition of certain anions where the order of effectiveness is EDTA > pyrophosphate > citrate ~ phosphate > F^- > tartrate > thiocyanate > sulphate. In each case, $Cr^{III}X$ ions are formed and the mechanism is thought to be

$$Cr^{II} + X \overset{\text{fast}}{\rightleftharpoons} Cr^{II}X$$

$$Cr^{II}X + Cr^{III} \rightleftharpoons Cr^{III}X + Cr^{II},$$

the second stage involving transfer of an electron. Similar effects are observed for the Cr^{2+}–$Co(NH_3)_5Cl^{2+}$ redox reaction, and, with the addition of pyrophosphate, both the Cl^- and $P_2O_7^{4-}$ are found in the inner sphere of the Cr^{III} product. This could only occur if the transition complex were

$$[P_2O_7Cr^{II}ClCo^{III}(NH_3)_5],$$

and it has been suggested that the role of the second anion in such cases is to raise the energy of the e_g electron to be transferred. Sulphate ions are less effective than $P_2O_7^{4-}$, and Cl^- ions (0·1 M) have no effect.

As can be seen in Table 14, *trans*-substituted Cr^{III} complexes react faster than the corresponding mono or *cis* complexes. This type of *trans*-effect can be accounted for in the following way. The incoming electron is accepted into an e_g orbital of Cr^{III} and the energy of the latter can be lowered by moving the groups *trans* to each other away from the metal ion. When one of the groups is a water molecule, as in $Cr(H_2O)_5Cl^{2+}$, stretching of the bond *trans* to the bridging group is more difficult than in the case of *trans*-$Cr(H_2O)_4Cl_2^+$, hence, electron transfer is more difficult.

The possibility of a transition complex with two ligands bridging the metal ions has been tested for in three different cases. In the first two instances, (*a*), the oxidation of Cr^{2+} by *cis*-$Co(NH_3)_4$

$(H_2O)_2^{3+}$, only one of the water molecules is transferred from the Co^{III} ion, and (*b*), in the oxidation of Cr^{2+} by *cis*-$Cr(H_2O)_4F_2^+$, monofluorochromium(III) is formed. More recently, however, (*c*), it has been shown that in the Cr^{2+} + *cis*-$Cr(H_2O)_4(N_3)_2^+$ reaction, both groups are transferred and *cis*-$Cr(H_2O)_4(N_3)_2^+$ is formed.[6] The only reasonable mechanism for the exchange is electron transfer via a transition complex,

$$\left[(H_2O)_4Cr^{II} \begin{matrix} \diagup N{-}N{-}N \diagdown \\ \diagdown N{-}N{-}N \diagup \end{matrix} Cr^{III}(H_2O)_4 \right]^{3+}$$

The rate constant $k = 60$ l mole^{-1} sec^{-1} is greater than $k =$ 1·2 l mole^{-1} sec^{-1} for the Cr^{2+}–$Cr(H_2O)_5N_3^{2+}$ reaction at 0°C.

TABLE 15

KINETIC DATA FOR Cr^{2+}–$Cr^{III}(NH_3)_5X^{2+}$ REACTIONS AT 25°C

Reactants	k (l mole^{-1} sec^{-1})	$\Delta S^{\ddagger}$ (e.u.)	$\Delta H^{\ddagger}$ (kcal mole^{-1})
$Cr^{2+} + Cr(NH_3)_5F^{2+}$	27×10^{-4}	−30	13·4
$Cr^{2+} + Cr(NH_3)_5Cl^{2+}$	$5{\cdot}7 \times 10^{-2}$	−23	11·1
$Cr^{2+} + Cr(NH_3)_5Br^{2+}$	0·32	−33	8·5
$Cr^{2+} + Cr(NH_3)_5I^{2+}$	5·5	—	—

The apparent lability of $Cr^{III}(NH_3)_5X^{2+}$ complexes with the addition of Cr^{2+} can be accounted for by a rate determining step

$$Cr(NH_3)_5X^{2+} + Cr^{2+} \rightarrow Cr(NH_3)_5H_2O^{2+} + CrX^{2+},$$

in which an electron and the X^- ligand are transferred. This is followed by the rapid hydrolysis of the labile Cr^{II} complex,

$$Cr(NH_3)_5H_2O^{2+} + 5H^+ \xrightarrow{\text{fast}} Cr^{2+} + 5NH_4^+.$$

6. R. SNELLGROVE and E. L. KING, *J. Am. Chem. Soc.* **84**, 4609 (1962).

Such reactions do not require the use of radioactive isotopes, since there is a net chemical change which can be followed spectrophotometrically. Rate constants (Table 15) increase as the polarizability of the bridging group increases, but other factors are probably involved since the entropies of activation do not vary in a similar manner.

The slowness of the exchange between $*Cr^{III}$ and Cr^{VI}, where the latter is present as the chromate ion CrO_4^{2-}, is to be expected first of all because of the difference in oxidation number, and secondly because the coordination number of the chromium is different in the two cases. The reaction has been studied at 94·8°C in 0·1 to 0·25 N acid.[7] At low Cr^{VI} concentrations the presence of bichromate can be neglected and the rate dependence is of the form $[Cr^{III}]^{4/3}[Cr^{VI}]^{2/3}$. On the basis of this rate law, it can be concluded that the rate determining step is the reaction of Cr^{III} with Cr^{V}, since the concentration of Cr^{V} is determined by the overall equilibrium

$$Cr^{III} + 2Cr^{VI} \rightleftharpoons 3Cr^{V},$$

and is therefore

$$[Cr^{V}] = K[Cr^{III}]^{1/3}[Cr^{VI}]^{2/3}.$$

The reaction sequence leading to exchange is

$$*Cr^{III} + Cr^{VI} \underset{\text{fast}}{\rightleftharpoons} *Cr^{IV} + Cr^{V} \quad (1)$$

$$*Cr^{III} + Cr^{V} \rightleftharpoons *Cr^{IV} + Cr^{IV} \quad (2)$$

$$*Cr^{V} + Cr^{VI} \overset{\text{fast}}{\rightleftharpoons} *Cr^{VI} + Cr^{V}, \quad (3)$$

there being kinetic evidence only for the existence of the highly reactive Cr^{IV} and Cr^{V} states. Because (1) and (3) are fast, it can be reasoned that Cr^{III} and Cr^{IV}, and Cr^{VI} and Cr^{V} have similar coordination numbers, while (2), involving a change in coordination number, is the slow step. Although the Cr^{IV} is, no

7. C. ALTMAN and E. L. KING, *J. Am. Chem. Soc.* **83**, 2825 (1961).

doubt, partly hydrolysed, it is thought to be essentially octahedral, while Cr^{V} is probably four coordinated and similar to the chromate ion.

Exchange reactions between different oxidation states of manganese

Normal exchange methods have indicated a rather fast exchange between Mn^{II} and Mn^{III}, the upper limit being 4 l mole^{-1} sec^{-1} at 25°C. More precise measurements are difficult since Mn^{III} readily disproportionates,

$$2Mn^{III} \rightleftharpoons Mn^{II} + Mn^{IV}, \; (K \sim 10^{-3}) \qquad (1)$$

with the precipitation of MnO_2. The precipitation of MnO_2 from 10^{-3} M solutions of Mn^{III} can be avoided by having a twenty-five-fold excess of Mn^{II} (to reduce the equilibrium concentration of Mn^{IV}) and ~3 N acid (to prevent precipitation of MnO_2); but in more recent studies, further difficulties have been encountered in that the exchange is dependent on the method used for separating the Mn^{II} from the Mn^{III}.[8] There are two possible paths for the exchange, the direct electron transfer,

$$*Mn^{II}(t_{2g}^3 e_g^2) + Mn^{III}(t_{2g}^3 e_g^1) \rightarrow Mn^{III}(t_{2g}^3 e_g^1) + Mn^{II}(t_{2g}^3 e_g^2), \qquad (2)$$

and the Mn^{III} disproportionation as in (1) above, in which case an $[Mn^{III}]^2$ dependence should be observed. Rates so far measured are faster than would be expected for a direct e_g electron transfer as in (2).

The Mn^{VI}–Mn^{VII} exchange between MnO_4^{2-} and MnO_4^- ions has been studied using radioactive tracers and NMR techniques, there being good agreement between the two.[9] An inner-sphere transition complex of the type

$$[O_3Mn—O—MnO_3]^-$$

can be excluded, since, in the presence of $H_2^{18}O$ solvent, there is

8. H. Diebler and N. Sutin, *J. Phys. Chem.* **68**, 174 (1964).
9. L. Gjertson and A. C. Wahl, *J. Am. Chem. Soc.* **81**, 1572 (1959); O. K. Myers and J. C. Sheppard, *J. Am. Chem. Soc.* **83**, 4716 (1961).

only slow exchange of oxygen atoms with MnO_4^{2-} and MnO_4^-. The exchange is furthermore catalysed by cations $Cs^+ > K^+ > Na^+ = Li^+$, and outersphere bridged transition complexes,

$$[O_3Mn—O—Cs—O—MnO_3]^{2-}$$

seem likely. At 0°C and in 0·16 N NaOH the second-order rate constant $k = 730$ l mole^{-1} sec^{-1}, while in 0·16 N CsOH $k = 2470$ l mole^{-1} sec^{-1}.

Exchange reactions between iron(II) *and iron*(III)

Although the Fe^{II}–Fe^{III} exchange has been extensively studied in the presence of a variety of anions,[(10)] details of the mechanism, whether inner or outer sphere, remain uncertain.† The inner co-ordination spheres of both the Fe^{II} and Fe^{III} aquo ions are readily substituted, and the exact composition of these for the product ions is therefore difficult to determine. The kinetic data is interpreted in terms of $Fe^{III}X^{2+}$ participation (Table 17) where the reaction sequence is believed to be

$$Fe^{3+} + X^- \overset{\text{fast}}{\rightleftharpoons} FeX^{2+}$$

$$FeX^{2+} + Fe^{2+} \xrightarrow{k}.$$

Evidence for the above has been obtained in the presence of fluoride ions ($X^- = F^-$).[(11)] By increasing the concentration of ferrous, it is possible to make the isotopic exchange rate comparable to the rate of formation of the inner coordination sphere

† Recent results (R. J. Campion, T. J. Conocchioli and N. Sutin, *J. Am. Chem. Soc.* **86**, 4591 (1964)) have shown that electron exchange between $FeCl^{2+}$ and Fe^{2+} proceeds mainly by a chloride-bridged inner-sphere complex. In addition, exchange can also proceed by an outer-sphere activated complex or a water-bridged inner-sphere activated complex, the latter accounting for the Fe^{2+} catalysed dissociation of the $FeCl^{2+}$ complex.

10. See, e.g. J. Silverman and R. W. Dodson, *J. Phys. Chem.* **56**, 846 (1952).

11. J. Menashi, S. Fukushima, C. Foxx and W. L. Reynolds, *Inorg. Chem.* **3**, 1242 (1964).

complex FeF^{2+}. Under these conditions, the exchange rate is found to be dependent on the rate of formation of FeF^{2+} in the manner expected if the activated complex containing one fluoride ion is formed from Fe^{2+} and FeF^{2+} (the reaction need not necessarily proceed by an inner-sphere mechanism however).

TABLE 16

KINETIC DATA FOR SOME Fe^{II}–Fe^{III} AQUO ION EXCHANGE REACTIONS AT 0°C IN PERCHLORATE MEDIA (μ = 0·55)

Reaction	k (l mole^{-1} sec^{-1})	$\Delta S^{\ddagger}$ (e.u.)	$\Delta H^{\ddagger}$ (kcal mole^{-1})
$Fe^{2+} + Fe^{3+}$	0·87	−25	9·3
$Fe^{2+} + FeOH^{2+}$	$1{\cdot}0 \times 10^3$	−18	6·9
$Fe^{2+} + FeF^{2+}$	9·7	−21	8·6
$Fe^{2+} + FeF_2^{+}$	2·5	−22	9·0
$Fe^{2+} + FeCl^{2+}$	5·4	−15	11·0
$Fe^{2+} + FeBr^{2+}$	4·9	−25	8·0
$Fe^{2+} + FeNCS^{2+}$	4·2	−21	9·2
$Fe^{2+} + FeN_3^{2+}$	$1{\cdot}9 \times 10^3$	7	13·3
(above 25°C)		<−20	<9·0
$Fe^{2+} + FeSO_4^{+}$	90	−19	9·0

The lowest rate is observed for the reaction between the hexaquo Fe^{3+} and hexaquo Fe^{2+} ions. Addition of a chloride ion to the activated complex increases the rate by a factor of about ten, but fluoride, chloride and bromide anions give approximately the same rate. If an inner-sphere bridging mechanism were operative, an order of effectiveness $Br^- > Cl^- > F^-$ would be expected, as in the Cr^{2+}–CrX^{2+} reactions. Because the activation parameters in Table 16 show little variation from one reaction to another, it has been suggested that some or all of these reactions proceed by a common mechanism in which there is hydrogen-atom bridging,

$$\left[\begin{array}{c} \quad\ \mathrm{H} \qquad\qquad\quad \mathrm{H} \\ \quad\ | \qquad\qquad\quad\ \ | \\ \mathrm{Fe^{II}}\!-\!\mathrm{O}\!-\!\mathrm{H}\!-\!-\!-\!\mathrm{O}\!-\!\mathrm{Fe^{III}}\!-\!\mathrm{X} \\ \qquad\qquad\qquad | \\ \qquad\qquad\qquad \mathrm{H} \end{array}\right]^{4+}$$

If this is so, it is unlikely that reactions proceed by a hydrogen-atom transfer mechanism (as some authors have suggested), since $FeOH_3^{3+}$ and $FeOH^{2+}$ are probably both endothermic, with respect to Fe^{2+} and Fe^{3+}. The enthalpy for the hydrolysis

$$Fe^{3+} + H_2O \rightleftharpoons FeOH^{2+} + H^+$$

is known to be 10 kcal mole^{-1}, so that activation energies of around 20 kcal mole^{-1} would be expected if this were, in fact, the mechanism. The symmetry of the complex in the Fe^{2+} + $FeOH^{2+}$ reaction

$$\left[\begin{array}{c} \quad\ \ H \qquad\quad\ \ H \\ \quad\ \ | \qquad\qquad | \\ Fe^{II}\text{—}O\text{—}H\text{---}O\text{—}Fe^{III} \end{array}\right]^{4+}$$

is thought to account for the enhanced rate of this path. A further possibility is that outer-sphere transition complexes of the type

$$\left[\begin{array}{c} Fe^{II}\text{—}O\text{—}H\text{---}X\text{—}Fe^{III} \\ | \quad \\ H \quad \end{array}\right]^{4+}$$

are formed.

The exchange between Fe^{II} and Fe^{III} has also been studied in a number of non-aqueous solvents. In both nitromethane and anhydrous alcohols, exchange is very slow, again suggesting that water plays a specific role in the reaction. That the constants for the Fe^{2+}–Fe^{3+}, Fe^{2+}–$FeOH^{2+}$ and Fe^{2+}–$FeCl^{2+}$ paths were lowered by a factor of two in D_2O, was thought to be further evidence for hydrogen-atom bridging. Doubts were expressed concerning the validity of this conclusion when a k_{H_2O}/k_{D_2O} ratio of 1·3 was found for the Cr^{2+}–$Co(NH_3)_5Cl^{2+}$ reaction, which is known to proceed by an inner-sphere chloride-bridged transition complex. Furthermore, the isotope effect on the equilibrium

$$Fe^{3+} + Cl^- \rightleftharpoons FeCl^{2+}$$

is of the same order of magnitude as the isotope effect on the Fe^{2+}–$FeCl^{2+}$ exchange, and hydrogen-atom bridging mechanisms are clearly not relevant in this equilibrium. More recently, equilibrium constants for the reaction

$$Fe^{3+} + D_2O \rightarrow FeOD^{2+} + D^+$$

have been measured.[12] These differ from those in H_2O, and activation parameters for the Fe^{2+}–$FeOD^{2+}$ reaction are found to be $\Delta H^\ddagger = 11{\cdot}5 \pm 0{\cdot}5$ kcal mole^{-1} and $\Delta S^\ddagger = -4{\cdot}0 \pm 1{\cdot}8$ e.u. Since these values differ so markedly from those for the Fe^{2+}–$FeOH^{2+}$ reaction, it might, in this one instance, be argued that they constitute evidence for an hydrogen-atom transfer mechanism. Until more is known of D_2O and H_2O solvent effects, the full implications of these results are difficult to assess.

The reaction of Fe^{2+} with FeN_3^{2+} differs from other Fe^{2+}–FeX^{2+} reactions in that the Arrhenius plot is curved above 13°C, and that, below 13°C, the activation parameters are numerically quite different from those for the other reactions in Table 16. A possible explanation of the curvature[13] is that electron transfer occurs within a binuclear-bridged complex $[Fe^{2+}—N_3—Fe^{3+}]$ and that, below 13°, the rate determining step is the formation of this complex while above 13°, the rate of electron transfer within the complex and the decomposition of the complex are comparable. In all other reactions one of the extremes seems to hold, that is, either electron transfer within the intermediate is much faster than its decomposition, or electron transfer is comparatively slow and the decomposition fast.

More recent work on the Fe^{2+}–Fe^{3+} exchange in a solid ice matrix is more difficult to interpret in terms of compact inner- or outer-sphere mechanisms. In the first place, the Arrhenius plot continues undeviated in going from the liquid to the solid phase, and secondly, since the Fe^{2+} and Fe^{3+} reactants would appear

---

12. S. Fukushima and W. L. Reynolds, *Talanta* **11**, 283 (1964).
13. D. Bunn, F. S. Dainton and S. Duckworth, *Trans. Farday Soc.* **57**, 1131 (1961).

to be immobile in the solid phase, it has been concluded that electron transfer takes place at distances of up to 100 Å.[14] If such a remote electron transfer is possible, one might ask how one reactant knows of the presence of the other; even at such remote distances, some degree of orbital overlap or link-up would seem necessary. The intermediate formation of solvated electrons is clearly not possible, owing to the form of the rate dependence (see p. 221).

When CN^-, phenanthroline and bipyridyl groups are co-ordinated to Fe^{II} and Fe^{III}, the field strength of the ligands is such that spin-paired t_{2g}^6 and t_{2g}^5 complexes are formed. That both $Fe(CN)_6^{4-}$ and $Fe(CN)_6^{3-}$ are, in fact, inert to substitution, can be confirmed by adding labelled cyanide ions to their respective solutions. The exchange between $Fe(CN)_6^{4-}$ and $Fe(CN)_6^{3-}$ can be studied in the presence of tetraphenylarsonium hydroxide and with $\sim 10^{-4}$ M EDTA to complex any simple cations which may be present as impurities. It is catalysed by the addition of H^+, K^+, Ca^{2+} and Ba^{2+}; with potassium ions, for example, there are terms in $[K^+]$ and $[K^+]^2$, and $KFe(CN)_6^{3-}$ and $K_2Fe(CN)_6^{2-}$ are probable reactants. Bridging of the two complexes by cations in the transition complex would seem likely.

Other outer-sphere reactions listed in the first part of Table 17 are considerably faster than Fe^{2+}–$Fe^{III}X^{2+}$ reactions already considered. This can be attributed to a number of factors, including (a), the greater similarity of metal–ligand bond distances in spin-paired $Fe^{II}(t_{2g}^6)$ and $Fe^{III}(t_{2g}^5)$ complexes, and (b), the increased conductivity of ligand groups. Ligands such as water, which have saturated single bonds, are not expected to conduct electrons between metal ions as readily as unsaturated ligand groups, such as CN^-. For those reactions in the second part of Table 17 in which there is a net chemical change, the favourable standard free energy ΔG^0 contributes to a lowering of the activation free energy $\Delta G^\ddagger$, where from transition-state theory,

$$\Delta G^\ddagger = -RT \log_e(kh/kT)$$

14. R. A. Horne, *J. Inorg. Nucl. Chem.* **25**, 1139 (1963).

i.e. k is bigger the more negative $\Delta G^{\ddagger}$ (see the section on theoretical considerations at the end of this chapter). For these reactions, a linear relationship between ΔG^{0} and $\Delta G^{\ddagger}$ (from $\Delta H^{\ddagger}$ and $\Delta S^{\ddagger}$ values) has been demonstrated.[15]

TABLE 17

KINETIC DATA FOR EXCHANGE REACTIONS INVOLVING SOME SPIN-PAIRED Fe^{II} AND Fe^{III} COMPLEXES

Reaction	Temp. (°C)	Rate constant ($l\ mole^{-1}\ sec^{-1}$)
$Fe(CN)_6{}^{4-} + Fe(CN)_6{}^{3-}$	0·1	355
$Fe(CN)_6{}^{4-} + Fe(phen)_3{}^{3+}$	25	$>10^8$
$Fe(phen)_3{}^{2+} + Fe(phen)_3{}^{3+}$	0	$>10^5$
$Fe(d\text{-}phen)_3{}^{2+} + Fe(d\text{-}phen)_3{}^{3+}$	0	$>2{\cdot}0 \times 10^3$
$Fe(d\text{-}bipy)_3{}^{2+} + Fe(phen)_3{}^{3+}$	25	$>10^8$
$Fe(t\text{-}phen)_3{}^{2+} + Fe(d\text{-}bipy)_3{}^{3+}$	25	$>10^8$
$Fe^{2+} + Fe(bipy)_3{}^{3+}$	25	$2{\cdot}7 \times 10^4$
$Fe^{2+} + Fe(phen)_3{}^{3+}$	25	$3{\cdot}7 \times 10^4$
$Fe^{2+} + Fe(5\text{-}Me\text{-}phen)_3{}^{3+}$	25	$2{\cdot}0 \times 10^4$
$Fe^{2+} + Fe(5\text{-}NO_2\text{-}phen)_3{}^{3+}$	25	$1{\cdot}1 \times 10^6$
$Fe^{2+} + Fe(5\text{-}Cl\text{-}phen)_3{}^{3+}$	25	$2{\cdot}1 \times 10^5$

bipy = 2,2′-bipyridine, d-bipy = 4,4′-dimethyl-2,2′-bipyridine
phen = 1,10-phenanthroline, t-phen = 3,4,7,8,-tetramethyl-1,10-phenanthroline.

Exchange reactions between cobalt(II) *and cobalt*(III)

Exchange reactions between $Co^{II}(t_{2g}^5 e_g^2)$ and $Co^{III}(t_{2g}^6)$ complexes are generally slow, (*a*) because e_g electrons are involved so that the difference in metal–ligand bond distances for the two forms is large, and (*b*) because the reaction is partially spin forbidden. By this we mean that direct electron transfer results in the formation of two excited states

$$Co^{II}(t_{2g}^5 e_g^2) + Co^{III}(t_{2g}^6) \rightarrow Co^{III}(t_{2g}^5 e_g^1) + Co^{II}(t_{2g}^6 e_g^1),$$

and further electronic changes must follow.

15. M. H. FORD-SMITH and N. SUTIN, *J. Am. Chem. Soc.* 83, 70 (1961).

In the first two reactions listed in Table 18, exchange is much faster than the above restrictions would seem to allow. The Co^{2+}–Co^{3+} exchange is thought to proceed by way of high-spin Co^{3+} since magnetic susceptibility measurements have shown that, in aqueous solution, a small fraction of the Co^{3+} is in the high-spin state.† That the high-spin state should be thermally accessible

$$Co^{3+}(t_{2g}^6) \overset{K}{\rightleftharpoons} Co^{3+}(t_{2g}^4 e_g^2),$$

TABLE 18

RATE CONSTANTS FOR SOME EXCHANGE REACTIONS BETWEEN Co^{II} AND Co^{III} COMPLEXES

Reaction	Temp. (°C)	Rate constant (l mole^{-1} sec^{-1})
$Co^{2+} + Co^{3+}$	25	~5·0
$Co(phen)_3{}^{2+} + Co(phen)_3{}^{3+}$	0	1·1
$Co(NH_3)_n{}^{2+} + Co(NH)_6{}^{3+}$	64·5	$<10^{-8}$
$Co(NH_3)_n{}^{2+}$ + *trans*-$Co(NH_3)_4(OH)_2{}^+$	64·5	$4·0 \times 10^{-3}$
$Co(NH_3)_n{}^{2+}$ + *cis*-$Co(NH_3)_4(OH)_2{}^+$	64·5	$2·5 \times 10^{-2}$
$Co(en)_n{}^{2+} + Co(en)_3{}^{3+}$	50	$1·4 \times 10^{-4}$
$Co(en)_n{}^{2+} + Co(en)_2(H_2O)_2{}^{2+}$	50	$2·7 \times 10^{-4}$
$Co(EDTA)^{2-} + Co(EDTA)^-$	100	$1·4 \times 10^{-4}$

is not unreasonable, since the water molecule has only a weak ligand field, and CoF_6^{3-} is known to be paramagnetic with four unpaired electrons. Providing the inter-conversion is fast and non-rate determining, such a path is in no way contrary to the second-order kinetics which are observed. Thus experimental rate constants will equal $k'K$ where k' is the true rate constant for the reaction

$$Co^{3+}(t_{2g}^4 e_g^2) + Co^{2+}(t_{2g}^5 e_g^2) \xrightarrow{k'}$$

† Accurate measurements are difficult, however, since Co^{3+} slowly oxidizes water and solutions are never completely free from Co^{2+}.

and K is the constant for the thermal equilibrium. The fast rate for the $Co(phen)_3^{2+}$–$Co(phen)_3^{3+}$ exchange can likewise be accounted for if a small fraction of the $Co(phen)_3^{2+}$ is present in the $t_{2g}^6 e_g^1$ spin-paired form[16]

$$Co^{2+}(t_{2g}^5 e_g^2) \rightleftharpoons Co^{2+}(t_{2g}^6 e_g^1)$$

$$Co^{3+}(t_{2g}^6) + Co^{2+}(t_{2g}^6 e_g^1) \rightarrow Co^{2+}(t_{2g}^6 e_g^1) + Co^{3+}(t_{2g}^6).$$

Again this is not unreasonable, since *o*-phenanthroline is known to have a high ligand field.

The slow rates which are observed for a number of redox reactions involving hexaquo cobalt(III) might also be explained by the intermediate formation of high-spin Co^{III} (present in only small amounts). The Co^{III} oxidation of Cr^{III} is, for example, much slower than the Ce^{IV} oxidation of Cr^{III} (although the oxidation potential of Co^{III} is higher),[17] while, from theoretical considerations, a rate constant, some 10^5 times that observed, would be expected for the reaction between Fe^{II} and Co^{III}.[18] Other factors may also be important, however. Thus, it has been suggested that in $2{\cdot}0 \times 10^{-3}$ M Co^{III} solutions, with the hydrogen-ion concentration $<0{\cdot}1$ M, a large proportion of the Co^{III} is present in the dimeric form (see page 207).

The most favourable path for the very much slower $Co(NH_3)_n^{2+}$–$Co(NH_3)_6^{3+}$ exchange, where $n = 3$ to 6, is thought to be one involving outer-sphere bridged transition complexes.[19] At 64·5°C and in aqueous ammonia solutions, pH ~ 10 the outer-sphere complex $Co(NH_3)_6^{3+}, OH^-$ is first formed

$$Co(NH_3)_6^{3+} + H_2O \overset{\text{fast}}{\rightleftharpoons} Co(NH_3)_6^{3+}, OH^- + H^+$$

---

16. B. R. Baker, F. Basolo and H. M. Neumann, *J. Phys. Chem.* **63**, 371 (1959).
17. J. B. Kirwin, P. J. Proll and L. H. Sutcliffe, *Trans. Faraday Soc.* **60**, 119 (1964).
18. G. Dulz and N. Sutin, *Inorg. Chem.* **2**, 917 (1963).
19. N. S. Biradar, D. R. Stranks and M. S. Vaidya, *Trans. Faraday Soc.* **58**, 2421 (1962).

where K is believed to be around 4×10^{-12} mole l^{-1}. This, then, reacts with a Co^{II} complex, e.g. $Co(NH_3)_6^{2+}$, giving a transition complex of the type

$$\left[\begin{array}{c} \quad\quad\quad\quad H \\ \quad\quad\quad\quad | \quad\quad\quad \leftarrow e \\ (NH_3)_5Co^{III}—N—H\ldots O\ldots Co^{II}(NH_3)_5 \\ \quad\quad\quad\quad | \quad\quad\quad\quad | \\ \quad\quad\quad\quad H \quad\quad\quad\quad H\rightarrow \end{array}\right]^{4+}$$

an electron and the hydroxyl group being transferred as indicated. Evidence for such a reaction path is as follows. Firstly, the initial distribution of labile cobalt(II) complexes $Co(NH_3)_6^{2+}$, $Co(NH_3)_5H_2O^{2+}$, $Co(NH_3)_4(H_2O)_2^{2+}$, etc., is different from that of the cobalt(III) ammines following exchange. The difference suggests that the conversion of Co^{II} to Co^{III} is accompanied by substitution of a hydroxide ion into the inner sphere of the newly formed Co^{III} complex. Secondly, the exchange shows an inverse acid dependence, which is in accordance with hydroxide ion participation. A consistent interpretation is possible if it is assumed that the $Co(NH_3)_n^{2+}$ complexes from $n = 3$ to $n = 6$ react at identical rates. At 64·5°C the rate constant for such $Co(NH_3)_n^{2+}$–$Co(NH_3)_6^{3+}$, OH^- reactions is $5{\cdot}6 \times 10^{-3}$ l mole^{-1} sec^{-1}. Chloride ions appear to function in a similar way, though net transfer cannot be detected since hydrolysis of cobalt(III) chloroammines is fast in basic solution. At 65·5°C the rate constant for the $Co(NH_3)_n^{2+}$–$Co(NH_3)_6^{3+}$, Cl^- reactions is $7{\cdot}3 \times 10^{-4}$ l mole^{-1} sec^{-1}. Similar anion-dependent paths are observed for the slow exchange between $Co(en)_n^{2+}$ and $Co(en)_3^{3+}$.

In the oxidation of $Co^{II}(CN)_5^{3-}$ by $Co^{III}(NH_3)_5X$ in the presence of free CN^-, inner- and outer-sphere mechanisms are indicated.[20] With $X = Cl^-$, N_3^-, NCS^- and OH^-, the rate law is of the form

$$\text{Rate} = k_1[Co^{II}(CN)_5^{3-}][Co^{III}(NH_3)_5X],$$

20. J. P. Candlin, J. Halpern and S. Nakamura, *J. Am. Chem. Soc.* **85**, 2517 (1963).

and since $Co^{III}(CN)_5X$ ions are formed, it is concluded that reaction proceeds with the formation of an inner-sphere transition complex,

$$[(CN)_5Co^{II}XCo^{III}(NH_3)_5].$$

TABLE 19

KINETIC DATA FOR INNER SPHERE (k_1) AND OUTER SPHERE (k_2) $Co^{II}(CN)_5{}^{3-}$–$Co^{III}(NH_3)_5X^{2+}$ REACTIONS AT 25°C, $\mu = 0·2$

X	k_1 (l mole^{-1} sec^{-1})	k_2 (l^2 mole^{-2} sec^{-1})
Cl^-	$\sim 5 \times 10^7$	—
N_3^-	$1·6 \times 10^6$	—
NCS^-	$1·1 \times 10^6$	—
OH^-	$9·3 \times 10^4$	—
NH_3	—	9×10^4
$SO_4{}^{2-}$	—	$3·6 \times 10^4$
OAc^-	—	$1·1 \times 10^4$
fumarate^{2-}	—	$1·2 \times 10^4$
oxalate^{2-}	—	$1·0 \times 10^4$
maleate^{2-}	—	$7·5 \times 10^3$
succinate^{2-}	—	6×10^3
$CO_3{}^{2-}$	—	$\sim 1 \times 10^3$
$PO_4{}^{3-}$	—	$5·2 \times 10^2$
F^-	$1·8 \times 10^3$	$1·7 \times 10^4$
NO_3^-	$\leqslant 1·0 \times 10^4$	$2·4 \times 10^5$

For a number of other groups, X = PO_4^{3-}, CO_3^{2-}, SO_4^{2-}, NH_3 and OAc^-, the reaction follows a different course and $Co^{III}(CN)_6^{3-}$ ions are formed. The rate law is

$$\text{Rate} = k_2[Co^{II}(CN)_5^{3-}][Co^{III}(NH_3)_5X][CN^-],$$

and reactions are thought to proceed by an outer-sphere mechanism involving $Co^{III}(NH_3)_5X$ and $Co^{II}(CN)_6^{4-}$, where the latter is presumed to exist in equilibrium with $Co^{II}[CN]_5^{3-}$,

$$Co^{II}(CN)_5^{3-} + CN^- \rightleftharpoons Co^{II}(CN)_6^{4-}.$$

With $Co(NH_3)_5F^{2+}$, $Co(NH_3)_5NO_3^{2+}$, both paths are observed, and these can be made to predominate in turn by varying the concentration of cyanide ions. The rate constants in Table 19 suggest that the outer-sphere reactions have a relatively small dependence on X^-.

The exchange between platinum(II) *and platinum*(IV)

Complexing anions play an essential role in exchange reactions between square-planar Pt^{II} and octahedral Pt^{IV} complexes.[21] The exchange has been most studied with chloride ions, when the rate law is of the general form

$$\text{Rate} = k_{obs}[Pt^{II}][Pt^{IV}][Cl^-].$$

In the exchange between $Pt(en)_2^{2+}$ and *trans*-$Pt(en)_2Cl_2^{2+}$ in the presence of free chloride, which is labelled, small equilibrium concentrations of the penta-coordinated Pt^{II} complex

$$Pt(en)_2^{2+} + {}^*Cl^- \underset{K}{\overset{\text{fast}}{\rightleftharpoons}} Pt(en){}^*Cl^+$$

are believed to be formed. The latter can then interact with the Pt^{IV} with the formation of a symmetrical transition complex

$$\left[\begin{matrix} & \text{en} & & \text{en} & \\ Cl^*\text{—} & Pt^{II} & \ldots Cl \ldots & Pt^{IV} & \text{—}Cl \\ & \text{en} & & \text{en} & \end{matrix}\right]^{3+},$$

and exchange can take place by the transfer of two electrons, the chloride-bridging group remaining attached to the newly formed Pt^{IV}

$$\begin{matrix} & \text{en} & & & \text{en} & & & & \text{en} & & & \text{en} & \\ Cl^*\text{—} & Pt^{II} & + & Cl\text{—} & Pt^{IV} & \text{—}Cl & \xrightarrow{k} & Cl^*\text{—} & Pt^{IV} & \text{—}Cl & + & Pt^{II} & \text{—}Cl. \\ & \text{en} & & & \text{en} & & & & \text{en} & & & \text{en} & \end{matrix}$$

The net result is, therefore, the substitution of a labelled chloride ion into the coordination sphere of the Pt^{IV}, the experimental rate constant k_{obs} being equal to kK. At 25°C, $k_{obs} = 2 \times 10^2$

21. F. Basolo, M. M. Morris and R. G. Pearson, *Discussions Faraday Soc.* **29**, 80 (1960).

l^2 mole^{-2} sec^{-1}. Direct exchange between free chloride ions and chloride complexed to Pt^{IV} is extremely slow. In support of the above mechanism there is no detectable exchange between *trans*-$Pt(tetrame\ en)_2Cl_2^{2+}$ and $Pt(tetrame\ en)_2^{2+}$, over fourteen days. This seems reasonable since the bulkiness of the tetramethylethylenediamine ligand, $NH_2 . C(CH_3)_2 . C(CH_3)_2 . NH_2$, will prevent a sufficiently close approach of the two reactants.

Reactions of the above type, in which there is a net chemical reaction, are very much influenced by the nature of the departing ligand. Thus, in the reaction between $Pt(NH_3)_4^{2+}$ and $Pt(NH_3)_5Cl^{3+}$,

$$Pt(NH_3)_4^{2+} + Cl^- \overset{\text{fast}}{\rightleftharpoons} Pt(NH_3)_4Cl^+$$

$$Pt(NH_3)_4Cl^+ + Pt(NH_3)_5Cl^{3+} \rightarrow Pt(NH_3)_4Cl_2^{2+} + Pt(NH_3)_4^{2+} + NH_3,$$

the rate is much slower since a Pt—NH_3 bond has to be broken.

The exchange between copper(I) *and copper*(II)

Cuprous ions readily disproportionate in aqueous perchlorate solution,

$$2Cu^+ \rightleftharpoons Cu^{2+} + Cu^0$$

and can exist as such in only minute equilibrium amounts. At high chloride-ion concentrations, the Cu^+ is stabilized by complexing, and in 12 N hydrochloric acid there is a fast exchange between Cu^I and Cu^{II} which has been measured using NMR line-broadening techniques ($k = 0{\cdot}5 \times 10^8$ l mole^{-1} sec^{-1}).[22] It has been suggested that a symmetrical chloride-bridged intermediate of the type

$$\left[\begin{matrix} Cl \\ Cl \end{matrix}\!\!>\!Cu{-}Cl{-}Cu\!<\!\!\begin{matrix} Cl \\ Cl \end{matrix}\right]^{2-}$$

is formed.

22. H. M. McConnell and H. E. Weaver, *J. Chem. Phys.* **25**, 307 (1956).

The exchange between silver(I) *and silver*(II)

In studying the exchange of Ag^{I} with Ag^{II}, fresh solutions of Ag^{II} have to be prepared for each set of experiments. This is conveniently done by dissolving AgO (from the anodic oxidation of $AgNO_3$) in 6 N perchloric acid. The Ag^{II} component of reaction solutions shows satisfactory stability for the duration of exchange experiments.

The rate law,

$$\text{Rate} = k[Ag^{II}]^2,$$

is independent of Ag^{I},[23] and is consistent with a disproportionation mechanism,

$$2Ag^{II} \rightleftharpoons Ag^{I} + Ag^{III},$$

the rate of exchange being equal to the rate of the disproportionation reaction at equilibrium. At 0°C and in 6 N perchloric acid, $k = 1 \times 10^3$ l mole^{-1} sec^{-1}, and since the equilibrium constant K is $\ll 1$, the rate constant for the reverse reaction, the reaction between Ag^{I} and Ag^{III}, must be large. The value of k is at least a hundred times greater than the rate constant for any direct electron transfer between Ag^{I} and Ag^{II}. The reaction has a high inverse dependence on the hydrogen-ion concentration, which suggests that the Ag^{II} species involved in the disproportionation are hydrolysed products such as $(AgOH)_2$ or possibly AgO.

B. LANTHANIDE AND ACTINIDE ELEMENTS

Exchange reactions between different oxidation states of the lanthanides

All of the lanthanides form stable 3+ ions, but the II and IV oxidation states, if they exist, tend to be highly reactive. As a result, only two reactions have so far been studied, those of Ce^{III} with Ce^{IV}, and Eu^{II} with Eu^{III}.

Even in strongly acid solution, the kinetics of Ce^{IV} reactions are complicated by the extensive hydrolysis of Ce^{4+}, the first

23. B. M. Gordon and A. C. Wahl, *J. Am. Chem. Soc.* **80**, 273 (1958).

hydrolysis constant being *ca.* 0·5 at 0°C.[24] The $CeOH^{3+}$ ions combine to form polynuclear ions and, in the exchange of Ce^{III} with Ce^{IV}, the rate equation and relevant exchanging species are believed to be

$$\text{Rate} = k_1[Ce^{3+}][Ce(OH)_2^{2+}] + k_2[Ce^{3+}][Ce(OH)_3^{+}] + k_3[Ce^{3+}][CeOCeOH^{5+}]$$

with smaller contributions from less hydrolysed forms. The reaction is strongly catalysed by fluoride and sulphate ions, but the effect of chloride is only slight and over long periods the chloride is oxidized to chlorine.

The exchange between Eu^{II} and Eu^{III} has been studied in a predominantly perchlorate medium.[25] With only perchlorate ions present the exchange is extremely slow, and over long periods the Eu^{II} tends to react with trace amounts of oxygen which are present (there is little or no reaction with perchlorate ions, see p. 208). The exchange has been studied with the addition of small amounts of chloride which are found to catalyse the reaction. By varying the amounts of chloride, it can be shown that the exchange is essentially first order in chloride and that there is little or no contribution from paths involving aquo or hydroxy ions (there is no intercept at zero chloride). It has been suggested that reaction is between Eu^{2+} and the chloro complex of Eu^{3+}

$$Eu^{2+} + EuCl^{2+} \xrightarrow{k}.$$

At 22°C the stability constant for the formation of $EuCl^{2+}$ is 0·9 l $mole^{-1}$, and $k \sim 2{\cdot}6 \times 10^{-4}$ l $mole^{-1}$ sec^{-1}.

The slowness of the Eu^{2+}—Eu^{3+} and Ce^{3+}—Ce^{4+} reactions is attributed to the smaller radial extension of *f* compared with *d* orbitals of similar energy.

---

24. F. R. Duke and F. R. Parchen, *J. Am. Chem. Soc.* **78**, 1540 (1956).
25. D. J. Meier and C. S. Garner, *J. Phys. Chem.* **56**, 853 (1952).

Actinide exchange reactions

In acid solution the various oxidation states of uranium (IV, V, and VI) and neptunium and plutonium (III, IV, V, and VI) are present as M^{3+}, M^{4+}, MO_2^+ and MO_2^{2+} ions respectively. As a result of the increased size of the 4+ ions, these are not as extensively hydrolysed as are V^{IV} and Ti^{IV} ions.

That the $M^{III} + M^{IV}$ and $M^V + M^{VI}$ exchange reactions are faster than the $M^{IV} - M^V$ reactions can be explained by considering the difference in degree of hydrolysis of the reactants. The exchange of Np^{4+} with NpO_2^+ is, for example, slow, and the half-life is *ca.* 12 hr at 50°C, while the exchange of UO_2^+ with UO_2^{2+} is fast, the rate constant being in the region 10^2–10^3 l mole^{-1} sec^{-1}. The corresponding reaction between NpO_2^+ and NpO_2^{2+} has been studied in some detail[26] and the rate may be expressed

$$\text{Rate} = k_1[NpO_2^+][NpO_2^{2+}] + k_2[NpO_2^+][NpO_2^{2+}][H^+].$$

At 4·5°C, $k_1 = 74$ l mole^{-1} sec^{-1}, and $k_2 = 15$ l^2 mole^{-2} sec^{-1}. The activation parameters for the acid independent path are $\Delta H^\ddagger = 11$ kcal mole^{-1} and $\Delta S^\ddagger = -12$ e.u. The second term is thought to correspond to the reaction of protonated NpO_2^+, i.e.

$$NpO_2H^{2+} + NpO_2^{2+} \rightarrow,$$

since the latter has the least positive charge. Transition complexes of the type

$$[O\text{—}Np^V\text{—}O \ldots H\text{—}O\text{—}H \ldots O\text{—}Np^{VI}\text{—}O]^{3+}$$

and

$$[O\text{—}Np^V\text{—}O\text{—}H \ldots O\text{—}Np^{VI}\text{—}O]^{4+}$$

have been suggested, but if hydrogen atoms play such a prominent part, a bigger D_2O solvent effect might be expected. For reactions in D_2O the decrease in the rate, $k_{H_2O}/k_{D_2O} = 1{\cdot}4$, is

---

26. J. C. Sullivan, D. Cohen and J. C. Hindman, *J. Am. Chem. Soc.* **79**, 3672 (1957).

little more than that for the Cr^{2+}–$Co(NH_3)_5Cl^{2+}$ inner-sphere reaction. Alternatively, an inner-sphere transition complex of the type

$$\left[\begin{array}{l} O \\ | \\ Np \ldots O—Np—O \\ | \\ O \end{array}\right]^{3+}$$

is not unreasonable. When chloride ions are added, further rate terms in $[Cl^-]$ and $[Cl^-]^2$ are obtained.

The rate equation for the Pu^{III}–Pu^{IV} exchange is

$$\text{Rate} = k_3[Pu^{3+}][Pu^{4+}] + k_4[Pu^{3+}][Pu^{4+}][H^+]^{-1},$$

and at 0°C, $k_3 \sim 200$ l mole^{-1} sec^{-1} and $k_4 \sim 1300$ sec^{-1}. The rapidity of the Pu^{3+}–Pu^{4+} path, which is some 10^2 times faster than the Fe^{2+}–Fe^{3+}, is a clear indication that charge repulsions are not of first importance in determining the rate of such electron-transfer processes.

Other actinide systems in which there is a net chemical change have also been studied. In, for example, the Np^{4+}–NpO_2^{2+} reaction,

$$Np^{IV} + Np^{VI} \rightarrow 2Np^{V},$$

the principal rate determining step is

$$-d[Np^{4+}]/dt = k_5[Np^{4+}][NpO_2^{2+}][H^+]^{-2},$$

with a much smaller contribution from a second term in $[H^+]^{-3}$.[27] At 25°C and $\mu = 2{\cdot}0$, $k_5 = 0{\cdot}045$ l^{-1}mole sec^{-1}. The equilibrium constant K for the overall reaction

$$Np^{4+} + NpO_2^{2+} + 2H_2O \rightleftharpoons 2NpO_2^{+} + 4H^{+},$$

in 1 N perchloric acid, and at 25°C is $5{\cdot}45 \times 10^6$ mole^4 l^{-4}, so

27. J. C. Hindman, J. C. Sullivan and D. Cohen, *J. Am. Chem. Soc.* **81**, 2317 (1959).

that k_{-5} for the Np^{V} disproportionation (which is given by k_5/K) is $9{\cdot}5 \times 10^{-9}$ l^3 $mole^{-3}$ sec^{-1}. The acid dependence for k_{-5} can also be deduced from that of k_5 and K. Thus

$$-d[NpO_2^+]dt = k_{-5}[NpO_2^+]^2[H^+]^2.$$

Why the acid dependence should be different from that for the U^{V} disproportionation,

$$d[UO_2^+]/dt = k_0[UO_2^+]^2[H^+],$$

is not readily explained.

C. NON-TRANSITION ELEMENTS

The exchange between mercury(I) *and mercury*(II)

The exchange between $(Hg^{I})_2$ and Hg^{II} is too fast to measure by normal isotopic labelling techniques. It is believed to proceed by the disproportionation reaction

$$(Hg^{I})_2 \rightleftharpoons Hg^0 + Hg^{II}.$$

The only alternative is the dissociation reaction

$$Hg_2^{2+} \rightarrow 2Hg^+,$$

followed by the rapid exchange of Hg^+ and Hg^{2+} ions. Evidence for the existence of $\sim 10^{-7}$ M concentrations of mercury atoms in aqueous solutions has been obtained by direct measurement of the solubility of mercury in water and by spectroscopic observations.

The kinetics of the $(Hg^{I})_2$–Mn^{III} and $(Hg^{I})_2$–Tl^{III} reactions are consistent with mechanisms in which there is an initial disproportionation of the $(Hg^{I})_2$, followed by oxidation of the mercury atoms so formed.

Exchange reactions between thallium(I) *and thallium*(III)

Of the exchange reactions between non-transition elements, that of Tl^{I} and Tl^{III} is the only one which can be satisfactorily studied in perchloric acid solutions, i.e. a non-complexing media. Hydrolysis of Tl^{3+},

$$Tl^{3+} + H_2O \rightleftharpoons TlOH^{2+} + H^+,$$

is more extensive than that of 3+ transition metal ions, the hydrolysis constant K being 0·07 mole l^{-1} at 25°C. The rate equation for the exchange reaction is of the form

$$\text{Rate} = k_1[Tl^+][Tl^{3+}] + k_2[Tl^+][TlOH^{2+}],$$

and at 25°C in perchlorate solutions $\mu = 3{\cdot}0$ M, $k_1 = 7{\cdot}03 \times 10^{-5}$ l mole^{-1} sec^{-1} and $k_2 = 2{\cdot}47 \times 10^{-5}$ l mole^{-1} sec^{-1}.[28] Thus, at this ionic strength, the overall rate decreases as the acidity is decreased and the presence of hydroxide ions in the transition complex produces a retardation. Although in 6 M perchlorate k_2 is somewhat bigger than k_1, the hydroxide ions clearly do not have the strong catalytic effect found in other metal ion reactions. On the other hand, it does not have the even stronger inhibitory effect which the chloride, bromide and cyanide can have.

For a series of runs with the addition of increasing amounts of chloride experimental second-order rate constants, k_{obs} first decrease and then increase as shown in Fig. 24. By using association constants for thallic chloride species, the shape of this curve can be accounted for by a rate equation,

$$\text{Rate} = k_0[Tl^+][Tl^{3+}] + k_3[Tl^+][TlCl^{2+}] + k_4[Tl^+][TlCl_4^-] + k_5[TlCl_2^-][TlCl_4^-] + k_6[TlCl_3][TlCl_4^-],$$

where k_0 is the rate constant in 1 M perchloric in the absence of chloride ions.[29] In this equation, $k_0 = 6{\cdot}6 \times 10^{-5}$, $k_3 = 0{\cdot}28 \times 10^{-5}$, $k_4 = 4{\cdot}58 \times 10^{-5}$, $k_5 = 1{\cdot}32$ and $k_6 = 18{\cdot}9$ l mole^{-1} sec^{-1}. Electron exchange only occurs, therefore, when the activated complex contains zero, one, four, six and seven

28. E. Roig and R. W. Dodson, *J. Phys. Chem.* **65**, 2175 (1961).
29. G. M. Waind, *Symposium on Coordination Chemistry*, Tihany, Hungary, 1964.

chloride ions. The rate of reduction of the monochloro complex is much less than that of the aquo complex, while $TlCl_2^+$, which exists over a large range of chloride concentrations is, within experimental accuracy, quite unreactive. The term in $k_4[Tl^+]$ $[TlCl_4^-]$ is not unambiguous in that exchange between TlCl and $TlCl_3$ is also possible. With six and seven chloride ions in the

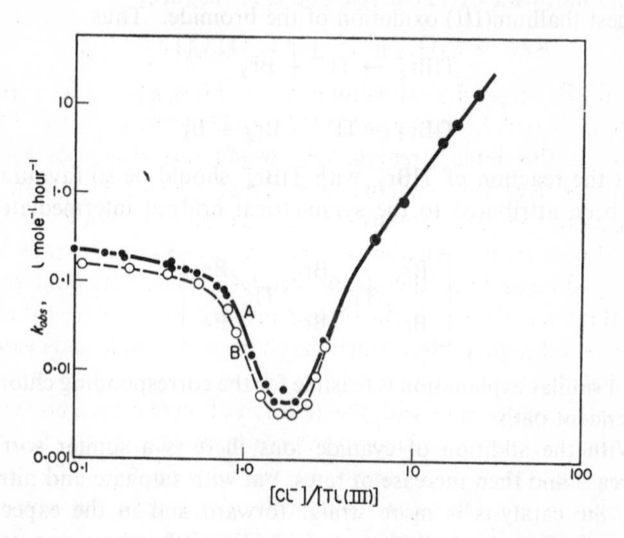

FIG. 24. The effect of chloride ions on the Tl^I–Tl^{III} exchange at 25°C, $[HClO_4]$ = 1·0 M, μ = 3·0 M. The broken line (curve B) is for the exchange in 80% D_2O. Logarithmic scales are used for convenience. [Reproduced, with permission, from S. GILKS, T. E. ROGERS and G. M. WAIND, *Trans. Faraday Soc.* **57**, 1373 (1961).]

transition complex, the effect of the chloride ions finally becomes catalytic. Since, at the higher chloride ion concentrations, there is no D_2O solvent effect, curve B in Fig. 24, the latter transition complexes at least are probably of the inner-sphere type.

For a similar series of experiments in which bromide ions are added in increasing amounts, the rate first decreases, increases to a maximum, falls to a second minimum and then increases again. A detailed investigation[30] has indicated an even more complex rate equation,

$$\text{Rate} = k_0[\text{Tl}^+][\text{Tl}^{3+}] + k_7[\text{TlBr}_2^+] + k_8[\text{TlBr}_3] + k_9[\text{Tl}^+][\text{TlBr}_4^-] + k_{10}[\text{TlBr}_2^-][\text{TlBr}_4^-].$$

The additional first-order terms here in thallium(III), k_7 and k_8, suggest thallium(III) oxidation of the bromide. Thus

$$\text{TlBr}_2^+ \rightarrow \text{Tl}^+ + \text{Br}_2$$

and

$$\text{TlBr}_3 \rightarrow \text{Tl}^+ + \text{Br}_2 + \text{Br}^-.$$

That the reaction of $TlBr_2^-$ with $TlBr_4^-$ should be so favourable has been attributed to the symmetrical bridged intermediate,

$$\left[\begin{matrix}\text{Br} & & \text{Br} & & \text{Br}\\ & \rangle\text{Tl}\langle & & \rangle\text{Tl}\langle & \\ \text{Br} & & \text{Br} & & \text{Br}\end{matrix}\right]^{2-}$$

and a similar explanation is feasible for the corresponding chloride dependent path.

With the addition of cyanide ions there is a similar sort of decrease and then increase in rates, but with sulphate and nitrate ions the catalysis is more straightforward and in the expected manner. From spectroscopic information it has been suggested that whereas chloride and bromide form inner-sphere complexes with Tl^{3+}, sulphate probably forms an outer-sphere complex $Tl(H_2O)_6^{3+},SO_4^{2-}$. On this evidence an outer-sphere transition complex,

$$[\text{Tl}^{\text{III}}\text{—H}_2\text{O—SO}_4\text{—Tl}^{\text{I}}]^{2+},$$

would seem likely. The rate equation also has an $[SO_4^{2-}]^3$ term and again it has been suggested that this path is more favourable than the $[SO_4^{2-}]^2$ path because a symmetrical transition complex can be formed.

---

30. L. G. Carpenter, M. H. Ford-Smith, R. P. Bell and R. W. Dodson, *Discussions Faraday Soc.* **29**, 92 (1960).

The kinetics of the exchange between Tl^{I} and Tl^{III} give no information concerning the possible intermediate formation of unstable Tl^{II}. That this is so can readily be appreciated, since both reaction sequences

$$(a) \qquad Tl^{I} + {}^{*}Tl^{III} \rightarrow Tl^{II} + {}^{*}Tl^{II}$$

$$Tl^{II} + {}^{*}Tl^{II} \xrightarrow{\text{fast}} Tl^{III} + {}^{*}Tl^{I},$$

$$\text{and } (b) \qquad Tl^{I} + {}^{*}Tl^{III} \rightarrow Tl^{III} + {}^{*}Tl^{I},$$

will give a rate equation which is first order in both $[Tl^{I}]$ and $[Tl^{III}]$. In a number of non-complementary reactions of the type

$$2Fe^{II} + Tl^{III} \rightarrow 2Fe^{III} + Tl^{I},$$

ample evidence has been obtained for the formation of Tl^{II} as a reaction intermediate. There has, therefore, been a good deal of speculation as to whether the Tl^{I}–Tl^{III} exchange is by successive one electron transfers with the intermediate formation of Tl^{II}, as in (*a*) above, or by simultaneous or near simultaneous transfer of two electrons as in (*b*). Evidence which has been obtained for a two electron change as in (*b*) is as follows[31]. The reaction

$$2V^{IV} + Tl^{III} \rightarrow 2V^{V} + Tl^{I},$$

has been studied at 80°C and is unaffected by the addition of excess Tl^{I} up to 0·13 M. Since the reaction of V^{IV} with Tl^{III} is believed to have a mechanism

$$V^{IV} + Tl^{III} \rightleftharpoons V^{V} + Tl^{II}$$

$$V^{IV} + Tl^{II} \rightarrow V^{V} + Tl^{I},$$

(the addition of V^{V} produces a marked retardation) and since the Tl^{I}–Tl^{III} exchange is known to proceed at an appreciable rate under the conditions of these experiments, it is concluded that the latter does not involve free Tl^{II}. If it did, then the progress of the reaction between V^{IV} and Tl^{III} would be affected.

---

31. A. G. SYKES, *J. Chem. Soc.* 5549 (1961).

A further distinction between the simultaneous transfer of two electrons and two successive one-electron transfers, within the lifetime of a *single* transition complex, is difficult to make. In certain instances, energetic considerations might exclude successive one electron transfers, an unfavourable free energy change for one of the steps being incompatible with the observed rate.

The exchange between tin(II) *and tin*(IV)

The choice of a suitable medium for studying the exchange of Sn^{II} with Sn^{IV} is somewhat restricted owing to the covalent nature of the Sn^{IV} state and its limited solubility in aqueous media. Solutions of hydrochloric acid are convenient, but in dilute acid a variety of hydrolysed and complexed ions are present and the nature of the reactants is uncertain. In 10 N hydrochloric acid, complexing is much more extensive and the exchanging species can be assumed to be $SnCl_4^{2-}$ and $SnCl_6^{2-}$.[32] It has not yet been established whether there is intermediate formation of the unstable Sn^{III} state.

For the exchange between $SnCl_2$ and $SnCl_4$ in methyl and ethyl alcohol,[33] the rate dependence is of the form $k[SnCl_2][SnCl_4]$ and activation energies are 20·9 and 23·7 kcal mole^{-1}, respectively. Both these values are considerably higher than the 10·8 kcal mole^{-1} for the reaction in 10 N hydrochloric acid, and a different mechanism would seem likely.

The exchange between antimony(III) *and antimony*(V)

The exchange of Sb^{III} with the Sb^{V} has been studied in hydrochloric acid solutions of up to 12 N. The complexity of the system at intermediate acid concentrations is largely due to the slow interconversion of two or more hydrolysed Sb^{V} species which exchange with Sb^{III} at different rates. In 9·5 N hydrochloric acid, the predominant form of Sb^{V} is $SbCl_6^-$ with a

32. C. I. Browne, R. P. Craig and N. Davidson, *J. Am. Chem. Soc.* **73**, 1946 (1951).
33. E. G. Meyer and A. Melnick, *J. Phys. Chem.* **61**, 367 (1957).

slight possibility that it is $HSbCl_6$,[34] while the most likely form of the Sb^{III} over the 1 N to 12 N hydrochloric acid range is thought to be $SbCl_4^-$ or $Sb(H_2O)_2Cl_4^-$.

There is no exchange in aqueous sulphuric acid solutions,[35] but this can be initiated by the addition of chloride ions. Exchange has been studied at two chloride-ion concentrations, 6 M and 0·2 M, where $[Sb^{III}]$, $[Sb^{V}]$, $[SO_4^{2-}]$ and $[H^+]$ are variables. In 6 M chloride, the sulphate ions have no effect and the exchange is the same as in hydrochloric acid solutions. At 0·2 M chloride, the system is more complex and both chloride and sulphate ions are involved in the activated complex.

In carbon tetrachloride, the exchange between $SbCl_3$ and $SbCl_5$ is zero order in $SbCl_3$, which implies a dissociation equilibrium,

$$SbCl_5 \rightleftharpoons SbCl_3 + Cl_2.$$

A similar dissociation constitutes the main path for the reaction of PCl_3 with PCl_5 in carbon tetrachloride.

The exchange between arsenic(III) *and arsenic*(V)

Exchange is measurable, though complex, in 10·8–12·6 M hydrochloric acid at 30°C.[36] Results indicate a dependence on the age of the As^V solutions and polymeric species may be involved. There is negligible exchange in 8·9 N hydrochloric acid over five days.

D. THEORETICAL CONSIDERATIONS

According to transition-state theory, the rate constant k for a particular reaction at a fixed temperature is defined by the activation parameters $\Delta H^\ddagger$ and $\Delta S^\ddagger$. Thus

$$k = (kT/h)e^{\Delta S^\ddagger/R}e^{-\Delta H^\ddagger/RT}$$

34. N. A. Bonner and W. Goishi, *J. Am. Chem. Soc.* **83**, 85 (1961).
35. J. A. Sincius, *Dissertation Abstr.* **23**, 431 (1962).
36. L. Le Van Anderson, *Dissertation Abstr.* **22**, 2598 (1962).

Factors which determine $\Delta H^{\ddagger}$ and $\Delta S^{\ddagger}$ (or more simply $\Delta G^{\ddagger}$, since $\Delta G^{\ddagger} = \Delta H^{\ddagger} - T\Delta S^{\ddagger}$) are, therefore, of some interest.[37]

For outer-sphere reactions of the non-bridged type, e.g. the majority of reactions between hexaquo ions, electron transfer is believed to occur by an electron tunnelling mechanism. By this we mean that electrons can, in effect, leak through a potential energy barrier which would classically be impenetrable. The result is that electrons are transferred at somewhat greater distances of upto say 10 Å, than would correspond to an actual collision of the reactants. The effect is clearly related to the extension of orbitals in space, though only a slight orbital overlap is possible.

According to the Franck–Condon principle, first proposed in connection with molecular spectra, electrons move much faster than nuclei, so that electronic transitions are complete before any significant nuclear changes take place. One of the factors opposing ready transfer of electrons is the difference in environment of the oxidized and reduced forms. Thus, in the reaction of Fe^{2+} with Fe^{3+}, the distance between the central metal atoms and coordinated water molecules differ by 0·15 Å. If electron transfer were to take place without any prior adjustment of metal–ligand bond distances, then two vibrationally excited ions would be formed, $Fe(H_2O)_6^{3+}$ with its Fe—O bonds more compressed, and $Fe(H_2O)_6^{3+}$ with its Fe—O bonds more extended than normal. For such a transfer to be possible, electronic activation energy would be required. If, alternatively, the metal–ligand bonds are first compressed or extended to a common distance, then electron transfer can take place without electronic excitation. If we consider the energy of the two ions immediately after electron transfer, it can be reasoned that the activation energy is significantly less in such a case than for a process in which there is electronic activation. Other factors which contribute to $\Delta H^{\ddagger}$ are the rearrangement of outer solvent molecules and the conductivity of ligand groups. Charge repulsions

37. J. Halpern, *Quart. Rev.* **15**, 228 (1961).

between reactants do not seem to be effective for outer-sphere reactions, and many reactions between highly charged ions of the same sign are fast, e.g. $Fe(CN)_6^{4-}$–$Fe(CN)_6^{3-}$ and $Fe(phen)_3^{2+}$–$Fe(phen)_3^{3+}$.

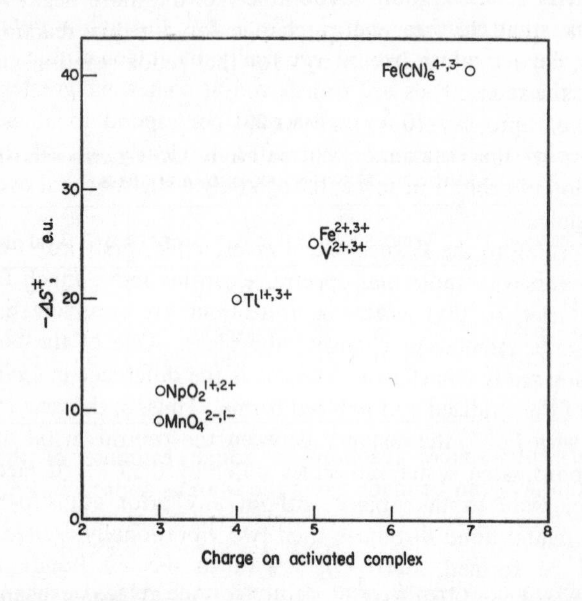

FIG. 25. The dependence of $\Delta S^{\ddagger}$ on the charge of the transition complex for some isotopic exchange reactions. [Reproduced, with permission, from J. HALPERN, *Quart. Rev.* **15**, 220 (1961).]

The entropy term for outer-sphere reactions is largely determined by solvation effects. Thus, for reactions between ions of the same charge type, negative entropy values are expected since the transition complex is more extensively solvated, and the system more ordered than the initial reactants. As the total charge on the transition complex is increased, $\Delta S^{\ddagger}$ becomes more negative, and for a number of systems a good linear relationship holds (Fig. 25).

Electron transfer reactions in which there is a net chemical change are generally much faster than the related exchange reactions; thus the Cr^{2+}–Fe^{3+} reaction is much faster than the Cr^{2+}–Cr^{3+} and Fe^{2+}–Fe^{3+} exchange reactions (Table 20). As a general rule, the more favourable (i.e. the more negative) the overall standard free energy change for a redox reaction, the lower the activation free energy for that redox reaction and the faster the rate.

TABLE 20

A COMPARISON OF KINETIC DATA FOR REDOX AND RELATED EXCHANGE REACTIONS BETWEEN HEXAQUO METAL IONS

Reaction	Temp. (°C)	k (l mole^{-1} sec^{-1})	$\Delta G^{\ddagger}$ (kcal mole^{-1})
Cr^{2+}–Fe^{3+}	25	$\sim 2 \times 10^{3}$	—
Cr^{2+}–Cr^{3+}	25	$\leqslant 2 \times 10^{-5}$	−24·4
Fe^{2+}–Fe^{3+}	0	0·87	−16·2
Fe^{2+}–Co^{3+}	0	10	−15·4
Co^{2+}–Co^{3+}	3·2	~ 1	−16·1

For outer-sphere reactions, a rough estimate of the rate constant k_{12} can be made using the Marcus approach. Equations of the type

$$k_{12} = (k_1 k_2 K_{12})^{1/2} \tag{1}$$

are obtained, where k_1 and k_2 are for the related exchange reactions, and K_{12} is the equilibrium constant for the redox reaction.[38] In a number of cases the order of magnitude of the estimated value is in good agreement with experimental values.[39] Furthermore, since

$$\Delta G^0_{12} = -RT \log_e K_{12},$$

and from transition state theory,

$$\Delta G^{\ddagger}_1 = -RT \log_e(k_1 h/ekT) \text{ and } \Delta G^{\ddagger}_2 = -RT \log_e(k_2 h/ekT)$$

it follows, from eqn. (1), that

$$\Delta G^{\ddagger}_{12} = \tfrac{1}{2}\Delta G^{\ddagger}_1 + \tfrac{1}{2}\Delta G^{\ddagger}_2 + \tfrac{1}{2}\Delta G^0_{12}.$$

38. See, e.g. N. SUTIN, *Ann. Rev. Nucl. Sci.* **12**, 308–25 (1962).
39. G. DULZ and N. SUTIN, *Inorg. Chem.* **2**, 917 (1963).

In effect, then, negative ΔG^0_{12} values serve to make $\Delta G^{\ddagger}_{12}$ less than might otherwise be expected from an average of $\Delta G^{\ddagger}_1$ and $\Delta G^{\ddagger}_2$.

For inner-sphere reactions, the theoretical approach is less well defined.[40] In forming the transition complex, there is significant orbital overlap and a new set of molecular orbitals are formed. Although these resemble closely those of the reactants, the electron distribution is such that, when the transition complex breaks down, there is a finite probability that an electron will be left behind. For inner-sphere reactions in which there is a symmetrical transition complex, the shape of molecular orbitals for the intermediate is probably such that electron transfer can take place much more readily. Some reorganization of ligand groups takes place in the formation of the transition complex, and the nature of the bridging group will obviously be rather critical. Because of the closer approach of the reactants, coulombic forces may be more important for inner-sphere reactions, especially if the bridging group is uncharged. The repulsion between cations is very much reduced by the presence of an anionic bridging group.

In general, the activation parameters for inner- and outer-sphere reactions do not differ sufficiently to allow a clear distinction between the two processes. Thus $\Delta H^{\ddagger}$ and $\Delta S^{\ddagger}$ values for the Fe^{2+}–Fe^{3+} exchange give no clear indication as to whether the reaction is inner or outer sphere. In the $Co(NH_3)_6^{2+}$–$Co(NH_3)_6^{3+}$ reaction, an outer-sphere mechanism in which there is hydroxide-ion bridging has been demonstrated, and the activation energy of $\sim$13 kcal mole^{-1} is considerably less than the value estimated for an electron tunnelling mechanism ($\sim$32 kcal mole^{-1}).

---

40. J. Halpern and L. E. Orgel, *Discussions Faraday Soc.* **29**, 32 (1960).

CHAPTER 8

REDOX REACTIONS BETWEEN METAL IONS IN AQUEOUS SOLUTIONS

FURTHER information regarding the closeness of approach of reactants in electron-transfer processes has been obtained by the study of redox systems. Not all these are simple one-stage processes, however, and, in many, the immediate interest has been one of resolving a complex reaction mechanism into a sequence of elementary steps. This is found most often (but not always) for reactions in which the stoicheiometric ratio of oxidant to reductant differs from unity, for example, the reaction of iron(II) with thallium(III),

$$2Fe^{II} + Tl^{III} \rightarrow 2Fe^{III} + Tl^{I},$$

in which Tl^{II} is formed as an intermediate. Again, unless otherwise stated, it should be assumed that reactions are in a perchlorate medium.

Some reactions of chromium(II) with cobalt(III) complexes

In the series of reactions of Cr^{2+} with $Co^{III}(NH_3)_5X^{2+}$ (Table 21) which are essentially oxidation–reduction reactions,

$$Cr^{II} + Co^{III} \rightarrow Cr^{III} + Co^{II},$$

transfer of the anion X^- to the newly formed Cr^{III} has been demonstrated. As in the Cr^{2+}–$Cr^{III}X^{2+}$ reactions, the only satisfactory explanation is that inner-sphere transition complexes,

$$[(H_2O)_5Cr^{II}—X—Co^{III}(NH_3)_5]^{4+},$$

are formed, the inner coordination spheres Co^{III} and Cr^{III} being inert to substitution, and those of Co^{II} and Cr^{II} labile.

TABLE 21

RATE CONSTANTS k_1 FOR THE REACTION OF SOME Co^{III} COMPLEXES WITH Cr^{II} (ASSUMED TO BE Cr^{2+}) AT 25°C. FOR REACTIONS IN WHICH X^- IS AN ORGANIC LIGAND WITH CONJUGATE BONDING, THERE IS AN ADDITIONAL ACID DEPENDENT TERM k_2

Reaction	k_1 (l mole^{-1}sec^{-1})	k_2 (l ^{2}mole^{-2}sec^{-2})
$Cr^{2+} + Co(NH_3)_5H_2O^{3+}$ (20°)	0·5	—
$Cr^{2+} + Co(NH_3)_5OH^{2+}$ (20°)	1·5 × 10^6	—
$Cr^{2+} + Co(NH_3)_5Cl^{2+}$	6 × 10^5	—
$Cr^{2+} + Co(NH_3)_5$(acetate)$^{2+}$	*a* 0·18	—
$Cr^{2+} + Co(NH_3)$ (H-succinate)$^{2+}$	*a* 0·17	—
$Cr^{2+} + Co(NH_3)_5$(benzoate)$^{2+}$	*a* 0·14	—
$Cr^{2+} + Co(NH_3)_5$ (H-fumarate)$^{2+}$	*b* 1·3	3·5
$Cr^{2+} + Co(NH_3)_5$ (H-maleate)$^{2+}$	*b* 200	100
$Cr^{2+} + Co(NH_3)_5$ (H-butadienedicarboxylate)	*b* 2·2	15·0
$Cr^{2+} + Co(NH_3)_5$(glycolate)	*c* 3·06	—
$Cr^{2+} + Co(NH_3)_5$(lactate)	*c* 6·65	—

a Adjacent attack; *b* Remote attack; *c* Adjacent attack with chelation.

The reactions in which X^- is a mono- or dicarboxylate group are of particular interest.[1] Let us consider, first of all, the case in which X^- is an acetate group, thus the Co^{III} reactant is

$$(NH_3)_5Co^{III}\text{—O—}\underset{\substack{\| \\ O}}{C}\text{—}CH_3.$$

Because the product is $Cr^{III}OOCCH_3$, Cr^{2+} attack must be at one of the oxygen atoms, and, until recently, the oxygen atom not attached directly to the cobalt was considered the more likely, since it is sterically less hindered. Experiments described below,

1. D. K. SEBERA and H. TAUBE, *J. Am. Chem. Soc.* **83**, 1785 (1961); R. D. BUTLER and H. TAUBE, *J. Am. Chem. Soc.* **87**, 5597 (1965); E. S. GOULD, *J. Am. Chem. Soc.* **87**, 4730 (1965).

in which $Co(C_2O_4)_3^{3-}$ is reduced with Cr^{2+}, have shown that, in this reaction at least, " alcoholic " as opposed to " ketonic " oxygen atoms may be involved, and that the bonding is

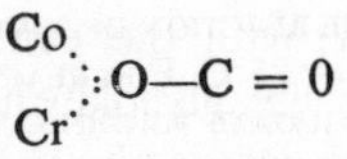

$$\begin{matrix} Co \\ Cr \end{matrix} \!\!\vdots O—C = 0$$

rather than Co...O$\cdots$C$\cdots$O...Cr in the transition complex. With a dicarboxylate group such as hydrogen succinate,

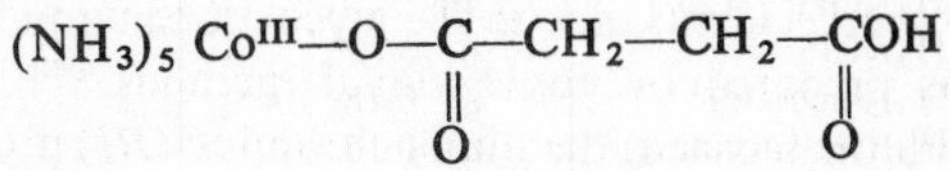

$$(NH_3)_5\,Co^{III}—O—\underset{\underset{O}{\|}}{C}—CH_2—CH_2—\underset{\underset{O}{\|}}{C}OH$$

the rate is similar to that for the acetate complex, and Cr^{2+} attack is probably at the carboxyl group adjacent to the Co^{III}. When the dicarboxylate group has a conjugate system of double bonds, Cr^{2+} attack at the carboxylate group at the remote end from that carrying the Co^{III} is thought to be favourable, the continuous pathway of double bonds providing an efficient means of electron transport between the two metal ions. Of the reactions listed in Table 21, the hydrogen fumarate, hydrogen maleate and hydrogen butadienedicarboxylate have conjugate bond systems. The reactions are dependent on the hydrogen-ion concentration:

Rate =

$$k_1[Cr^{2+}][Co(NH)_3)_5X^{2+}] + k_2[Cr^{2+}][Co(NH_3)_5X^{2+}][H^+].$$

Thus, with the hydrogen fumarate complex, for example, a transition complex

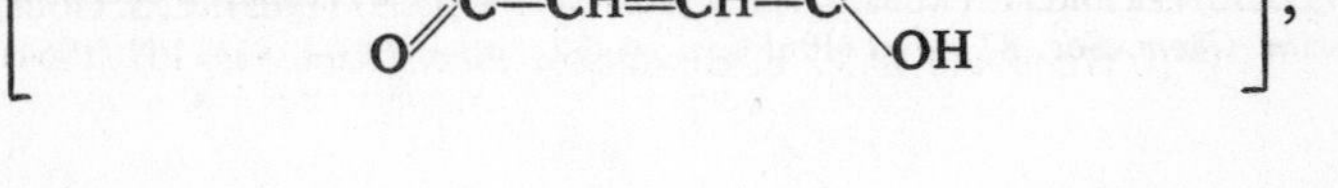

$$\left[(NH_3)_5Co^{III}—O\!\!\!\!\!\!\underset{O}{>}C—CH{=}CH—C\!\!\!\!\!\!\underset{OH}{<}^{O—Cr^{II}(H_2O)_5} \right]^{4+},$$

is formed, and the positioning of an additional proton on the carbonyl group adjacent to the Co^{III} is thought to improve conjugation.

The effect of chelation of the reducing agent is illustrated by the reaction of chromium(II) with the pentamminecobalt(*III*) complexes of α-hydroxy acids, e.g. glycolate and lactate. With glycolate as the bridging ligand the rate is about 16 times faster than when it is acetate. This is consistent with the view that chelation of the Cr^{2+} by glycolate takes place in the activated complex. The initial chromium(*III*) product is metastable and reverts (half-life ~ 22 hours) to a species identical with the glycolatopentaquochromium (*III*) ion as prepared by conventional methods. This evidence together with the fact that the initial chromium(*III*) product has a higher extinction coefficient indicates that the chromium is chelated by the bridging group in the activated complex. The rate is independent of the hydrogen-ion concentration over the range 0·05–1·0 N.

In the reaction between Cr^{2+} and $Co(C_2O_4)_3^{3-}$, the nature of the products has been determined by means of ion-exchange chromatography, and by measuring absorption spectra in the visible.[2] At hydrogen-ion concentrations below 0·02 M, at least 90% of the Cr^{III} is attached to three carboxylate groups, the product $Cr(H_2O)_3(C_2O_4)O_2CCO_2^-$ containing one bidentate and one unidentate oxalate group. At higher hydrogen-ion concentrations, only one of the oxalato groups is transferred, and above 0·2 N acid, 90% or more of the product is in the form $Cr(H_2O)_4(C_2O_4)^+$. The acid effect is presumably due to a protonation of the complex $Co(C_2O_4)_3^{3-}$. If, at the lower hydrogen-ion concentration, transfer of all three carboxylate groups occurs simultaneously, then the transition complex must be triply bridged. A model of the transition complex shows that Cr^{2+} can only attack oxygen atoms adjacent to the Co^{III}, the other oxygen atoms being too far apart. Three carboxylate groups are also transferred in the reaction between Cr^{2+} and

2. P. B. Wood and W. C. E. Higginson, *Proc. Chem. Soc.* 109 (1964).

Co^{III} EDTA at hydrogen-ion concentrations below 0·01 M.

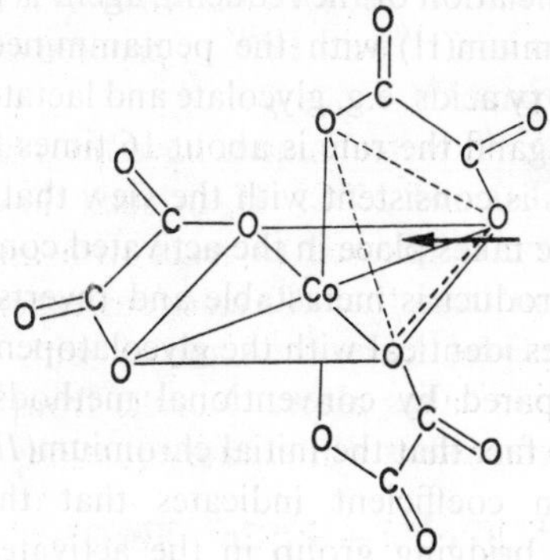

A comparison has been made of the reduction of some $Co^{III}(NH_3)_5X^{2+}$ complexes by Cr^{2+}, V^{2+} and Eu^{2+}, using, in most cases, the stopped-flow method.[3] The variation in rate with V^{2+} as reductant was found to be much smaller than that of Cr^{2+}, and to show sufficient similarity to the reactivity pattern for $Cr(bipy)_3^{2+}$ as to suggest that both V^{2+} and $Cr(bipy)_3^{2+}$ react by outer-sphere mechanisms. In contrast, the rates for the reactions of Eu^{2+} with halogenopentamminecobalt(III) complexes are more in keeping with an inner-sphere mechanism.

The reaction of chromium(II) with iron(III)

At room temperature, the reaction of Cr^{II} with Fe^{III} is too fast to be followed using normal techniques. In the presence of halide ions, X^-, non-labile $Cr^{III}X^{2+}$ complexes are formed, but since both reactants are labile it is not immediately clear which of

3. J. P. Candlin, J. Halpern and D. L. Trimm, *J. Am. Chem. Soc.* **86**, 1019 (1964).

them brings the X^- group into the transition complex. The Fe^{III} is less labile than the Cr^{II}, and at $-50°C$ in 5·27 N perchloric acid, spectrophotometric measurements have shown that the formation of $FeCl^{2+}$ is not sufficiently fast for the chloride-catalysed reaction of Cr^{II} with Fe^{III} to proceed exclusively by the $Cr^{2+} + FeCl^{2+}$ path.[4]

More recently,[5] flow techniques have been used to study the reaction at $\mu = 1{\cdot}0$ and 25°C. If, in such flow experiments, a Cr^{2+}–Cl^- solution is allowed to react with an Fe^{3+} solution, the rate is different from that of analogous experiments in which both cation solutions have been mixed with Cl^- before the oxidation–reduction takes place. Results from a series of such experiments are consistent with a rate equation,

$$\text{Rate} = k_1[Cr^{2+}][Fe^{3+}] + k_2[Cr^{2+}][FeOH^{2+}] + k_3[Cr^{2+}][FeCl^{2+}] + k_4[Cr^{2+}][Fe^{3+}][Cl^-].$$

The individual rate constants being $k_1 = 2{\cdot}3 \times 10^3$ l mole^{-1} sec^{-1}, $k_2 = 3{\cdot}3 \times 10^6$ l mole^{-1} sec^{-1}, $k_3 = 2{\cdot}0 \times 10^7$ l mole^{-1} sec^{-1} and $k_4 = 2{\cdot}0 \times 10^4$ l^2 mole^{-2} sec^{-1}. Both chloride-dependent paths yield $CrCl^{2+}$, so that k_3, at least, must be by an inner-sphere mechanism. The nature of the second chloride-dependent path, k_4, is less certain. Two formulations which are consistent with the observed rate law are

$$Cr(H_2O)_6^{2+} + Fe(H_2O)_6^{3+}, Cl^- \rightarrow [(H_2O)_5Cr^{II}\text{—}Cl\text{—}H_2O\text{—}Fe^{III}(H_2O)_5]^{4+} \rightarrow Cr(H_2O)_5Cl^{2+} + Fe(H_2O)_6^{2+}$$

and

$$Cr(H_2O)_5Cl^{+} + Fe(H_2O)_6^{3+} \rightarrow [(H_2O)_5Cr^{II}\text{—}Cl\text{—}H_2O\text{—}Fe^{III}(H_2O)_5]^{4+} \rightarrow Cr(H_2O)_5Cl^{2+} + Fe(H_2O)_6^{2+}.$$

4. M. Ardon, J. Levitan and H. Taube, *J. Am. Chem. Soc.* **84**, 872 (1962).
5. G. Dulz and N. Sutin, *J. Am. Chem. Soc.* **86**, 829 (1964).

It is not possible to distinguish between these on the basis of evidence at present available. In order to account for the formation of $CrCl^{2+}$, the first must proceed via an outer-sphere activated complex in which the chloride is bonded directly to the chromium. The second may also involve an outer-sphere transition complex, but the further possibility of a water bridged inner-sphere transition complex,

$$[(H_2O)_4ClCr^{II}—H_2O—Fe^{III}(H_2O)_5]^{4+},$$

TABLE 22

RELATIVE RATES FOR SOME REACTIONS OF Cr^{2+} WITH $Fe^{III}X^{2+}$, Fe^{2+} WITH $Fe^{III}X^{2+}$ AND Cr^{2+} WITH $Cr^{III}X^{2+}$

X^-	Cr^{2+}–$Fe^{III}X$	Cr^{2+}–$Cr^{III}X^{2+}$	Fe^{2+}–$Fe^{III}X^{2+}$
H_2O	1	1	1
OH^-	$1{\cdot}4 \times 10^3$	$>3{\cdot}5 \times 10^4$	$1{\cdot}16 \times 10^3$
Cl^-	1×10^4	$>4{\cdot}5 \times 10^5$	11·1
Br^-	—	$>3{\cdot}0 \times 10^6$	11·1

with the chloride acting as a non-bridging ligand, cannot be entirely ruled out. The latter seems unlikely, since, for reactions which are known to proceed by an inner-sphere mechanism, e.g. the Cr^{II}–$Co^{III}(NH_3)_5X^{2+}$ reactions, water is not a particularly efficient bridging group.

When relative rates for k_1, k_2 and k_3 are compared with those for the reactions of Fe^{2+} with $Fe^{III}X^{2+}$ and Cr^{2+} with $Cr^{III}X^{2+}$, Table 22, they are seen to resemble, closely, the latter. Reactions between hexaquo ions are generally assumed to be outer sphere, while the reactions of Cr^{2+} with $Cr^{III}X^{2+}$, in which $X^- =$ OH^-, Cl^- and Br^-, are known to proceed by an inner-sphere mechanism. It seems safe to assume that other Cr^{2+} reactions with $Fe^{III}X^{2+}$ proceed, at least in part, by an inner-sphere mechanism, but an inner-sphere mechanism would seem less likely for reactions of Fe^{2+} with $Fe^{III}X^{2+}$.

Some reactions between iron (II) and cobalt(III)

The reaction of Fe^{II} with Co^{III} has been studied by a quenching technique.[6] In this method, samples of reaction solution are injected into an alkaline α,α'-bipyridine solution and the absorbancy of $Fe(bipy)_3^{2+}$ measured. The rate equation is of the form

$$\text{Rate} = k_1[Fe^{2+}][Co^{3+}] + k_2[Fe^{2+}][CoOH^{2+}],$$

and at 0°C ($\mu = 1{\cdot}0$), k_1 and k_2 have values 10 and 6500 l mole^{-1} sec^{-1}, respectively. The corresponding $\Delta H^{\ddagger}$ values are 9·1 and 7·9 kcal mole^{-1}, and $\Delta S^{\ddagger}$ values are -23 and -14 entropy units, respectively.

Because of the large difference in the oxidation potentials of the Fe^{2+}/Fe^{3+} and the Co^{2+}/Co^{3+} couples, a faster rate might have been expected (i.e. the rate is some 10^5 times slower than that predicted by the Marcus Theory). The reaction probably proceeds with intermediate formation of spin-free Co^{3+}, but this cannot be rate determining since, if the reaction sequence were

$$Co^{3+}(t_{2g}^6) \rightleftharpoons Co^{3+}(t_{2g}^4 e_g^2)$$

$$Co^{3+}(t_{2g}^4 e_g^2) + Fe^{2+} \xrightarrow{\text{fast}} Co^{2+} + Fe^{3+},$$

the kinetics would show zero-order dependence on the ferrous.

Using flow techniques similar to those for the reaction of Cr^{II} with Fe^{III}, it has been established that, with the chloro complex $CoCl^{2+}$ as oxidant, the reaction proceeds by an inner-sphere mechanism.[7]

$$CoCl^{2+} + Fe^{2+} \rightarrow [CoClFe]^{4+} \rightarrow Co^{2+} + FeCl^{2+}.$$

In these experiments, $FeCl^{2+}$ concentrations forty times in excess of the final equilibrium concentrations, were detected during the reaction. On complexing with ammonia, the oxidation potential of Co^{III} is considerably reduced, and reactions between

6. L. E. Bennett and J. C. Sheppard, *J. Phys. Chem.* **66**, 1275 (1962).
7. T. J. Conocchioli, G. H. Nancollas and N. Sutin, *J. Am. Chem. Soc.* **86**, 1453 (1964).

Fe^{2+} and Co^{III} pentammine complexes are found to be slow. The rate of the reaction between Fe^{2+} and $Co(NH_3)_5N_3^{2+}$ has been compared with the rates of the corresponding Fe^{2+}–*cis*-$Co(NH_3)_4(N_3)_2^+$ and Fe^{2+}–*trans*-$Co(NH_3)_4(N_3)_2^+$ reactions.[8] All three show a first-order dependence on each of the reactants, the rate equations being of the form,

$$-d[Fe^{2+}]/dt = (k_1 + k_2[H^+])[Fe^{2+}][Co^{III}],$$

TABLE 23

RATE CONSTANTS k_1 AND k_2 FOR THE REACTION OF Fe^{2+} WITH SOME AZIDE COMPLEXES OF COBALT(III) AT 25°C

Co^{III} complex	k_1 ($l\ mole^{-1}\ sec^{-1}$)	k_2 ($l^2\ mole^{-2}\ sec^{-1}$)
$Co(NH_3)_5N_3{}^{2+}$	0·009	—
trans-$Co(NH_3)_4(N_3)_2{}^+$	0·073	1·36
cis-$Co(NH_3)_4(N_3)_2{}^+$	0·186	—

with k_2 equal to zero except in the case of the *trans* complex. Assuming that reactions proceed with the formation of azide-bridged transition complexes, the *trans* complex would be expected to react the fastest as in the case of the Cr^{2+} + *trans*-$Cr^{III}Cl_2^+$ reaction. Results show that, while the *trans* complex reacts faster than the mono-azide complex (Table 23), k_1 for the *cis* complex is faster still. The most probable explanation is that the *cis* complex forms a dibridged transition complex,

$$\left[(H_2O)_4Fe^{II}\begin{matrix} N{=}N{=}N \\ N{=}N{=}N \end{matrix}Co^{III}(NH_3)_4\right]^{3+},$$

in which case, a direct acid dependence would not be expected. The *trans* complex, on the other hand, can form a transition complex

$$[(H_2O)_5Fe^{II}{-}N{=}N{=}N{-}Co^{III}(NH_3)_4N_3]^{3+}$$

8. A. HAIM, *J. Am. Chem. Soc.* 85, 1016 (1963).

in which a proton can attach itself to the non-bridging azide group and in this way facilitate electron transfer.

Some reactions of titanium(III)

Perchlorate ion oxidation of the titanium(III) hexaquo ion is at such a rate that it is difficult to prepare stock solutions free from appreciable amounts of chloride (the reaction with perchlorate, here, is some 10^2 times faster than with V^{II}, and 10^3 times faster than with V^{III}). To determine the effectiveness of the chloride, redox reactions can be studied in a perchlorate medium in the presence of known amounts of chloride. Rate constants for the chloride independent reaction can then be obtained by extrapolating to zero chloride.

The effect which chloride ions have on Ti^{III} reactions depends a great deal on the nature of the oxidizing ion. Thus, in the reaction of Ti^{III} with Fe^{III}, the rate is first order in both reactants,

$$-d[Ti^{III}]/dt = k_{obs}[Ti^{III}][Fe^{III}],$$

where k_{obs} is dependent on both the chloride- and hydrogen-ion concentrations,

$$k_{obs} = k_1 + k_2[H^+]^{-1} + k_3[Cl^-]$$

A similar first-order dependence on metal-ion concentrations is observed in the reaction between Ti^{III} and Hg^{II}.[9] The first step

$$Ti^{III} + Hg^{II} \rightarrow Ti^{IV} + Hg^{I},$$

is rate determining and is presumably followed by

$$Ti^{III} + Hg^{I} \xrightarrow{\text{fast}} Ti^{IV} + Hg^{0}$$

and

$$Hg^{0} + Hg^{II} \xrightarrow{\text{fast}} (Hg^{I})_2.$$

The hydrogen-ion dependence for the rate determing step may be expressed

$$k'_{obs} = k_4[H^+]^{-1} + k_5[H^+]^{-2},$$

9. R. Critchley and W. C. E. Higginson, Private communication.

which is not unreasonable since both Ti^{3+} and Hg^{2+} are fairly extensively hydrolysed and Ti^{IV} is present as either oxo or hydroxo forms. Chloride ions have been shown to retard the reaction. It is of interest to compare the similar inhibitory effect which chloride ions have on the reactions of Fe^{II} with Tl^{III} and Tl^{I} with Tl^{III}, since Tl^{III} is isoelectronic with Hg^{II}, and both monochloro complexes have covalent character.

In the reaction of titanium(III) and plutonium(IV),

$$Ti^{III} + Pu^{IV} \rightarrow Ti^{IV} + Pu^{III},$$

the rate equation is of the form

$$-d[Ti^{III}]/dt = k_9[Ti^{III}][Pu^{IV}][H^+]^{-1},$$

and chloride ions have little effect on the rate.[10]

Some reactions of 4 + *actinide ions*

A feature of reactions in which the extent of hydrolysis of one of the reactants increases markedly in going to products is the large inverse acid dependence. Thus in the iron(III) oxidation of neptunium(IV) to neptunium(V),[11] overall reaction

$$Np^{4+} + Fe^{3+} + 2H_2O \rightarrow NpO_2^+ + Fe^{2+} + 4H^+,$$

the rate law is

$$-d[Np^{IV}]/dt = k[Np^{IV}][Fe^{III}][H^+]^{-3}.$$

Direct loss of an electron from Np^{4+} would probably require an oxidizing agent of exceptionally high electron affinity. Instead, with Fe^{III}, reaction is more favourable when the hydrolysis of Np^{4+} approaches that of the product.

In the reaction between uranium(IV) and iron(III),[12] overall equation

$$U^{4+} + 2Fe^{3+} + 2H_2O \rightarrow UO_2^{2+} + 2Fe^{2+} + 4H^+,$$

10. S. W. Rabidean and R. J. Kline, *J. Phys. Chem.* **64**, 193 (1960).
11. L. B. Magnusson and J. R. Huizenga, *J. Am. Chem. Soc.* **73**, 3202 (1951).
12. R. H. Betts, *Can. J. Chem.* **33**, 1780 (1955).

U^{V} is formed in the rate determining step. It is, perhaps, surprising that the acid dependence,

$$-d[U^{IV}]/dt = [U^{IV}][Fe^{III}](k_1[H^+]^{-1} + k_2[H^+]^{-2})$$

does not correspond to that for the reaction of Np^{IV} with Fe^{III}, since U^{V} is also present as the singly-charged cation UO_2^+.

The greater hydrogen-ion dependence of the reaction between U^{IV} and Ce^{IV},

$$-d[U^{IV}]/dt = [U^{IV}][Ce^{IV}](k_1[H^+]^{-2} + k_2[H^+]^{-3})$$

is probably due to the more extensive hydrolysis of Ce^{IV}.

The formation of thallium(II) in some non-complementary reactions

Because stable oxidation states of iron differ by one electron and those of thallium ($6s^26p^1$) by two, the reaction between Fe^{II} and Tl^{III} is said to be non-complementary. The stoicheiometric equation is

$$2Fe^{II} + Tl^{III} \rightarrow 2Fe^{III} + Tl^{I},$$

and since termolecular reactions are unlikely, an unstable intermediate, either Fe^{IV} or Tl^{II}, must be formed. When the products Fe^{III} and Tl^{I} are added, Fe^{III} but not Tl^{I} retards the reaction, which is in accord with a mechanism,

$$Fe^{II} + Tl^{III} \underset{k_{-1}}{\overset{k_1}{\rightleftharpoons}} Fe^{III} + Tl^{II}$$

$$Fe^{II} + Tl^{II} \xrightarrow{k_2} Fe^{III} + Tl^{I}.$$

With no Fe^{III} added, initially, second-order plots are linear until sufficient Fe^{III} has accumulated, i.e. at about 60% completion, when the back reaction k_{-1} becomes effective and there is curvature. From the full kinetic treatment (see p. 50) k_1 and k_2/k_{-1} can be determined. The variation of k_1 with the hydrogen-ion concentration suggests two paths for reaction, one with a single hydroxide ion and the other with two hydroxide ions in the transition complex. Since hydrolysis of Fe^{2+} is small, it is generally assumed that the relevant reactions are

$$Fe^{2+} + TlOH^{2+} \rightarrow$$

and

$$Fe^{2+} + Tl(OH)_2^+ \rightarrow .$$

The ratio k_2/k_{-1} is also acid dependent, but actual reaction paths are much less certain. Small amounts of chloride retard the reaction due to the formation of unreactive $TlCl^{2+}$, but larger amounts produce a catalytic effect as in the exchange of Tl^{I} with Tl^{III}.

An interesting catalytic effect is produced by introducing a piece of platinum foil into a stirred reaction solution. The platinum catalyses reaction k_2 with the result that k_{-1} can be ignored and the slow step k_1 becomes the dominant rate determining step. Under these conditions, second-order plots are linear to completion and k_1 can be measured.

Thallium(II) is also formed as an intermediate in reactions in which the Tl^{I} oxidation state is oxidized. Thus, the reaction of Tl^{I} with Co^{III}[13] proceeds by a mechanism

$$Tl^{I} + Co^{III} \rightleftharpoons Tl^{II} + Co^{II}$$

$$Tl^{II} + Co^{III} \rightarrow Tl^{III} + Co^{II}.$$

At 53·9°C and in 6·2 N nitric acid reaction of Tl^{I} with Ce^{IV} follows a similar reaction sequence,

$$Tl^{I} + Ce^{IV} \rightleftharpoons Tl^{II} + Ce^{III} \quad (1)$$

$$Tl^{II} + Ce^{IV} \rightarrow Tl^{III} + Ce^{III}, \quad (2)$$

but there are additional reactions involving hydroxyl radicals,[14]

$$Ce^{IV}OH^- \rightarrow Ce^{III} + OH \quad (3)$$

$$Tl^{I} + OH \rightarrow Tl^{II} + OH^-. \quad (4)$$

The Tl^{II} ions from (4) react with Ce^{III} and Ce^{IV} as in (1) and (2).

---

13. K. G. Ashurst and W. C. E. Higginson, *J. Chem. Soc.* 343 (1956).
14. M. K. Dorfman and J. W. Gryder, *Inorg. Chem.* **1**, 799 (1962).

Some reactions of the mercury(I) Dimer

Reactions in which $(Hg^{I})_2$ ions are oxidized to Hg^{II} can proceed by two entirely different paths depending on the nature of the metal ion used as oxidizing agent.

In the reaction with cobalt(III),

$$(Hg^{I})_2 + 2Co^{III} \rightarrow 2Hg^{II} + 2Co^{III},$$

oxidation is by the more direct route, with intermediate formation of Hg^{I}. Thus, the rate equation is first order in both reactants, which is in agreement with a reaction sequence,

$$(Hg^{I})_2 + Co^{III} \rightarrow Hg^{I} + Hg^{II} + Co^{II}$$

$$Hg^{I} + Co^{III} \xrightarrow{\text{fast}} Hg^{II} + Co^{II}.$$

The reaction of $(Hg^{I})_2$ with Ag^{II}

$$(Hg^{I})_2 + 2Ag^{II} \rightarrow 2Hg^{II} + 2Ag^{I},$$

is thought to proceed in a similar manner, since the kinetics of the reaction of $(Hg^{I})_2$ with Ce^{IV} in the presence of Ag^{I} as catalyst[15] are consistent with a first step

$$Ag^{I} + Ce^{IV} \rightarrow Ag^{II} + Ce^{III},$$

followed by

$$(Hg^{I})_2 + Ag^{II} \rightarrow Hg^{I} + Hg^{II} + Ag^{I}$$

and

$$Hg^{I} + Ce^{IV} \xrightarrow{\text{fast}} Hg^{II} + Ce^{III}.$$

In the reactions with manganese(III) and thallium(III), on the other hand, the oxidation proceeds by an indirect route involving Hg^{0}. The latter is present in minute quantities as a result of the rapid disproportionation

$$(Hg^{I})_2 \overset{K_1}{\rightleftharpoons} Hg^{0} + Hg^{II},$$

---

15. W. C. E. Higginson, D. R. Rosseinsky, B. Stead and A. G. Sykes, *Discussions Faraday Soc.* **29**, 49 (1960).

the equilibrium constant K_1 being of the order 10^{-9} l mole^{-1}. Thus in the reaction of $(Hg^{I})_2$ with Tl^{III},

$$(Hg^{I})_2 + Tl^{III} \rightarrow 2Hg^{II} + Tl^{I},$$

the rate determining step is believed to be the electron transfer

$$Hg^{0} + Tl^{III} \xrightarrow{k_2} Hg^{II} + Tl^{I}.$$

The rate equation is therefore

$$-d[Tl^{III}]/dt = k_2[Hg^{0}][Tl^{III}],$$

and substituting for $[Hg^{0}]$ from the pre-equilibrium, (see p. 53),

$$\frac{-d[Tl^{III}]}{dt} = \frac{k_2 K_1[(Hg^{I})_2][Tl^{III}]}{[Hg^{II}]},$$

which, in agreement with experiment, i.e. the addition of Hg^{II} slows the reaction down. The alternative mechanism

$$(Hg^{I})_2 + Tl^{III} \rightarrow 2Hg^{II} + Tl^{I},$$

would be expected to give rise to a simpler rate equation, while mechanisms involving Tl^{II} would almost certainly give more complex rate equations. The rate constant k_2 for the reaction of Hg^{0} with $TlOH^{2+}$ is 10^6 at 25°C.

The tendency of Mn^{III} to disproportionate,

$$2Mn^{III} \rightleftharpoons Mn^{II} + Mn^{IV},$$

with the precipitation of manganese dioxide, can be countered by having a twenty-five fold excess of Mn^{II}, and acid concentrations of around 3 N. The kinetics of the reaction between $(Hg^{I})_2$ and Mn^{III},[16]

$$(Hg^{I})_2 + 2Mn^{III} \rightarrow 2Hg^{II} + 2Mn^{II},$$

---

16. D. R. Rosseinsky, *J. Chem. Soc.* 1181 (1963).

are consistent with a mechanism

$$Mn^{III} + Hg^{0} \xrightarrow{k_3} Mn^{II} + Hg^{I}$$

$$Mn^{III} + Hg^{I} \xrightarrow{\text{fast}} Mn^{II} + Hg^{II}.$$

In addition Mn^{IV} ions from the disproportionation of Mn^{III} react directly with $(Hg^{I})_2$,

$$Mn^{IV} + (Hg^{I})_2 \rightarrow Mn^{II} + 2Hg^{II}.$$

The rate constant k_3 is 5×10^3 l mole^{-1} sec^{-1} at 50°C.

Summarizing, then, with sufficiently strong oxidizing agents such as Co^{III} (oxidation potential 1·84 V) and Ag^{II} (1·98 V), oxidation of $(Hg^{I})_2$ proceeds by the more direct route with the formation of Hg^{II} and Hg^{I} in the rate determining first step. With less strong oxidizing agents such as Mn^{III} (1·51 V) and Tl^{III} (1·25 V), reaction proceeds by disproportionation of $(Hg^{I})_2$ and subsequent oxidation of Hg^{0}. It is perhaps surprising that both paths are not effective with the stronger oxidizing agents, i.e. with Co^{III} and Ag^{II}, and some small (but so far undetected) contributions to the rate might have been expected.

The reaction between chromium(III) and cerium(IV)

The oxidation of Cr^{III} by Ce^{IV},

$$3Ce^{IV} + Cr^{III} \rightarrow 3Ce^{III} + Cr^{VI},$$

is too fast to be studied by conventional techniques in perchlorate solutions, but proceeds at a convenient rate in acid sulphate solutions.[17] The reaction can be followed spectrophotometrically at 500 mμ, Cr^{III} having the strongest absorption at this wavelength. Dimerization of Ce^{IV} is suppressed by the presence of the sulphate ions, while monomeric and dimeric forms of Cr^{VI} have approximately equal absorptions (per g atom of Cr^{VI}) at

17. J. Ying-Pek Tong and E. L. King, *J. Am. Chem. Soc.* **82**, 3805 (1960).

500 mμ. Since back reactions involving Cr^{VI} are not effective, it is immaterial whether the Cr^{VI} is in the monomeric or dimeric state. At 25°C the rate law is of the form

$$\frac{d[Cr^{VI}]}{dt} = \frac{k[Cr^{III}][Ce^{IV}]^2}{[Ce^{III}]},$$

which is consistent with a mechanism

$$Cr^{III} + Ce^{IV} \underset{k_{-1}}{\overset{k_1}{\rightleftharpoons}} Cr^{IV} + Ce^{III} \text{ (fast)} \quad (1)$$

$$Cr^{IV} + Ce^{IV} \overset{k_2}{\rightarrow} Cr^{V} + Ce^{III} \quad (2)$$

$$Cr^{V} + Ce^{IV} \rightarrow Cr^{VI} + Ce^{III} \text{ (fast).} \quad (3)$$

In these reactions, the Cr^{IV} and Cr^{V} oxidation states are formed as highly reactive intermediates. Small equilibrium concentrations of Cr^{IV} are produced in the rapid equilibrium first step

$$[Cr^{IV}] = \frac{k_1}{k_{-1}} \frac{[Cr^{III}][Ce^{IV}]}{[Ce^{III}]},$$

and react subsequently with Ce^{IV} in the rate determining step. The rate equation is therefore

$$\frac{d[Cr^{VI}]}{dt} = \frac{k_2 k_1}{k_1} \frac{[Cr^{III}][Ce^{IV}]^2}{[Ce^{III}]},$$

which is in agreement with experiment. That reaction (2) is rate determining suggests that the conversion of Cr^{IV} to Cr^{V} involves a change in coordination number from six to four. Similar conclusions were possible in the exchange between Cr^{III} and Cr^{VI}.

The reactions of chromium(VI) with vanadium(IV) and iron(II)

The kinetics of the oxidation of vanadium(IV) with chromium(VI) have been studied spectrophotometrically with reaction solutions 0·005 N in perchloric acid and $\mu = 1{\cdot}0$ M.[18] Under

18. J. H. Espenson, *J. Am. Chem. Soc.* **86**, 1883 (1964).

these conditions, the Cr^{VI} is predominantly in the form of the mononuclear hydrogen chromate ion, $HCrO_4^-$, and the overall reaction may be expressed

$$3VO^{2+} + HCrO_4^- + H^+ \rightarrow 3VO_2^+ + Cr^{3+} + H_2O.$$

The rate equation is of the form

$$\frac{-d[HCrO_4^-]}{dt} = \frac{k[VO^{2+}]^2[HCrO_4^-]}{[VO_2^+]},$$

which is in agreement with a mechanism

$$V^{IV} + Cr^{VI} \underset{k_{-1}}{\overset{k_1}{\rightleftharpoons}} V^V + Cr^V \text{ (fast)}$$

$$V^{IV} + Cr^V \xrightarrow{k_2} V^V + Cr^{IV}$$

$$V^{IV} + Cr^{IV} \rightarrow V^V + Cr^{III} \text{ (fast)},$$

the second stage corresponding to the Cr^V to Cr^{IV} change, being rate determining. Thus the rate may be expressed

$$\frac{-d[Cr^{VI}]}{dt} = k_2[V^{IV}][Cr^V],$$

and on substituting for Cr^V from the rapid equilibrium first step

$$\frac{-d[Cr^{VI}]}{dt} = \frac{k_1 k_2}{k_{-1}} \frac{[V^{IV}]^2[Cr^{VI}]}{[V^V]}.$$

At 25°C, $k = k_1 k_2 / k_{-1} = 0{\cdot}62$ l mole^{-1} sec^{-1}.

A similar rate law and mechanism has been found for the reaction of Fe^{II} with Cr^{VI}, but this system is more complex in that the interaction of Fe^{III} with Cr^{VI}

$$Fe^{3+} + HCrO_4^- \rightleftharpoons FeCrO_4^- + H^+$$

has to be taken into account.

The reaction between vanadium(III) and iron(III)

The reaction of V^{III} with Fe^{III} conforms to a stoicheiometric equation

$$V^{III} + Fe^{III} \rightarrow V^{IV} + Fe^{II},$$

and was thought, at first, to be a single-stage process in accordance with this equation. When, however, quantities of V^{IV} were added in an attempt to detect the back reaction, the rate was found to increase not decrease (Fig. 26). The effect of the

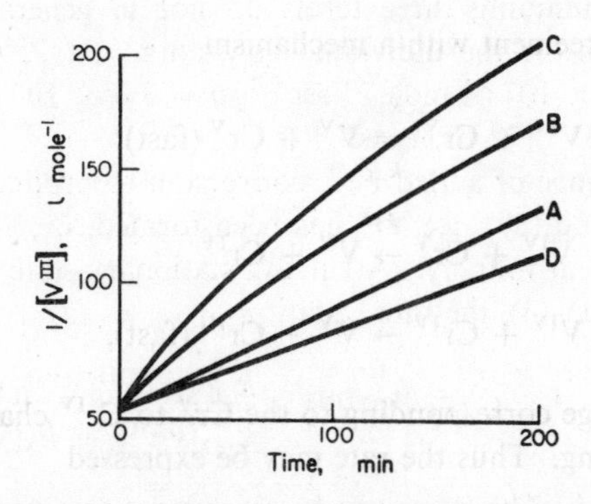

FIG. 26. Second-order plots for the reaction of V^{III} with Fe^{III} (both reactants 0·022 M). Curves B and C show the effect of 0·057 M and 0·094 M V^{IV} added initially. D shows the effect of quantities of Fe^{II} greater than 0·02 M. [Reproduced, with permission, from W. C. E. HIGGINSON and A. G. SYKES, *J. Chem. Soc.* 2841 (1962).]

V^{IV} was countered by adding Fe^{II} which is in agreement with a reaction sequence

$$V^{III} + Fe^{III} \xrightarrow{k_1} V^{IV} + Fe^{II} \qquad (1)$$

$$V^{IV} + Fe^{III} \underset{k_{-2}}{\overset{k_2}{\rightleftharpoons}} V^{V} + Fe^{II} \qquad (2)$$

$$V^{V} + V^{III} \xrightarrow{k_3} 2V^{IV}. \qquad (3)$$

Although the equilibrium in (2) lies well to the left (i.e. $K_2 \sim 10^{-6}$ at 0°C), reaction (3) is sufficiently fast to make the route provided by reactions (2) and (3) effective in the absence of large amounts of Fe^{II}. On adding excess of Fe^{II}, the concentration of V^{V} is so reduced that (2) and (3) make negligible contribution to the rate. Under these conditions k_1 can be obtained directly from second-order plots (curve D in Fig. 26), and from a series of such runs at different acid concentrations k_1 may be expressed,

$$k_1 = a + b[H^+]^{-1} + c[H^+]^{-2}.$$

Although k_1 values are subject to relatively small errors, such equations containing three terms do not in general give very precise values for the individual constants. At 25°C and $\mu = 3{\cdot}0$, $a = 2{\cdot}2 \times 10^{-3}$ l mole^{-1} sec^{-1}, $b = 3{\cdot}0 \times 10^{-3}$ sec^{-1} and $c = 3{\cdot}4 \times 10^{-3}$ l^{-1} mole sec^{-1}.

In the absence of added Fe^{II}, conversion to products can proceed by (2) and (3) once V^{IV} has been formed, or, alternatively, if V^{IV} is present initially. Assuming stationary-state kinetics for V^{V} in eqns. (1)–(3), the rate equation

$$\frac{-d[V^{III}]}{dt} = k_1[Fe^{III}][V^{III}] + \frac{k_2k_3[Fe^{III}][V^{III}][V^{IV}]}{k_{-2}[Fe^{II}] + k_3[V^{III}]},$$

can be derived. The latter can be rearranged to give

$$\frac{[Fe^{III}][V^{IV}]}{-d[V^{III}]/dt - k_1[Fe^{III}][V^{III}]} = \frac{k_{-2}[Fe^{II}]}{k_2k_3[V^{III}]} + \frac{1}{k_2},$$

and since k_1 is known, the left-hand side of this equation can be plotted against $[Fe^{II}]/[V^{III}]$. Such plots appear to pass through the origin, and $1/k_2$ is therefore small, but the gradient which is equal to k_{-2}/k_2k_3 can be measured. To obtain k_3 from the latter, the equilibrium constant k_2/k_{-2} can be determined under similar conditions of ionic strength and temperature by measuring the e.m.f. of V^{IV}/V^{V} and Fe^{II}/Fe^{III} half-cells at known acid concentrations. At 25°C and at 1 N hydrogen-ion concentration, $k_3 = 250$ l mole^{-1} sec^{-1}. There is good agreement between k_3 values and those obtained more recently from the direct study of the reaction of V^{III} with V^{V}.

The V^{III}–Fe^{III} reaction is catalysed by Cu^{II}.[19] With Cu^{II} concentration at least equivalent to those of Fe^{III}, the uncatalysed path may be ignored, and the rate

$$d[V^{III}]/dt = k_{cat}[V^{III}][Cu^{II}],$$

is independent of $[Fe^{III}]$. This is in agreement with a reaction sequence

$$V^{III} + Cu^{II} \xrightarrow{k_{cat}} V^{IV} + Cu^{I}$$

$$Cu^{I} + Fe^{III} \xrightarrow{\text{fast}} Cu^{II} + Fe^{II},$$

so that, under these conditions, information can be obtained regarding the reaction of V^{III} with Cu^{II}. The rate constant k_{cat} shows a dependence on $[H^+]$ in solutions, $\mu = 3{\cdot}0$ M ($NaClO_4$).

$$k_{cat} = f + g[H^+]^{-1}.$$

In 1 N perchloric acid and at 25°C $k_{cat} = 0{\cdot}30$ l mole^{-1} sec^{-1}. Compare $k_1 = 0{\cdot}025$ l mole^{-1} sec^{-1} for the one-stage reaction of V^{III} with Fe^{III} under similar conditions. That Cu^{II} is so effective as a catalyst is, in this instance, due to the much more favourable entropy of activation than the activation energy. Thus, if we consider the acid-independent paths a and f, the activation parameters are $\Delta H_a^\ddagger = 17{\cdot}3 \pm 4{\cdot}1$ kcal mole^{-1} and $\Delta S_a^\ddagger = -15 \pm 13$ e.u., while $\Delta H_f^\ddagger = 21{\cdot}3$ kcal mole^{-1} and $\Delta S_f^\ddagger = 5 \pm 5$ e.u. Note that $\Delta H_f^\ddagger$ for the catalysed reaction is actually greater than $\Delta H_a^\ddagger$ for the uncatalysed reaction.

The reaction between iron(II) and vanadium(V)

The fairly rapid reaction between Fe^{II} and V^{V},

$$Fe^{II} + V^{V} \rightarrow Fe^{III} + V^{IV},$$

has been studied by an extension of ordinary spectrophotometric

19. W. C. E. Higginson and A. G. Sykes, *J. Chem. Soc.* 2841 (1962).

methods,[20] and also by polarographic techniques (see p. 34). In the former, reactions were started by injecting the required amount of Fe^{II} into the V^{V} solution in a thermostatted spectrophotometric cell, and, at the same time, a spectrophotometric recorder chart was started. Optical densities were recorded at a wavelength in the 300–350 mμ region, where V^{V} has a strong absorption. The reaction shows a first-order dependence on both reactants,

$$-d[V^{V}]/dt = k_{obs}[Fe^{II}][V^{V}],$$

the acid dependence being of the form

$$k_{obs} = a[H^{+}]^{-1} + b + c[H^{+}].$$

It has been suggested that path a is due to reaction of the hydroxy form of VO_2^+, i.e. HVO_3, and c to reaction of a protonated form of VO_2^+. At room temperature, a is small and cannot be detected, but at 55°C ($\mu = 1{\cdot}0$), $a = 17$ sec^{-1}, $b = 240$ l mole^{-1} sec^{-1} and $c = 4610$ l^2 mole^{-2} sec^{-1}. That c is so prominent is reasonable, since in reducing VO_2^+ to VO^{++} there is a decrease in hydrolysis, which protonation will tend to favour. The overall activation energy for path c, $\Delta H^{\ddagger} = 1{\cdot}52$ kcal mole^{-1}, is surprisingly low compared with the corresponding value of 8·6 kcal mole^{-1} in the reaction of Fe^{II} with Np^{V}, where Np^{V} is present as NpO_2^+.

The reaction between vanadium(III) and cobalt(III)

The mechanism of the reaction of V^{III} with Co^{III}[21] is similar to that of the reaction of V^{III} with Fe^{III}. Thus

$$V^{III} + Co^{III} \xrightarrow{k_1} V^{IV} + Co^{II} \quad (1)$$

$$V^{IV} + Co^{III} \xrightarrow{k_2} V^{V} + Co^{II} \quad (2)$$

$$V^{III} + V^{V} \xrightarrow{\text{fast}} 2V^{IV}. \quad (3)$$

20. N. A. Daugherty and T. W. Newton, *J. Phys. Chem.* **67**, 1090 (1963).
21. D. R. Rosseinsky and W. C. E. Higginson, *J. Chem. Soc.* 31 (1960).

In eqn. (2), the equilibrium lies well to the right and the back reaction is not therefore effective. The rate constant k_2 can be obtained by separate study of the reaction between V^{IV} and Co^{III} and is used to evaluate k_1 from the rate equation

$$\frac{-d[Co^{III}]}{dt} = k_1[V^{III}][Co^{III}] + k_2[V^{IV}][Co^{III}].$$

TABLE 24

A COMPARISON OF RATE CONSTANTS ($l\ mole^{-1}\ sec^{-1}$) AT 0°C AND IN 1 N PERCHLORIC ACID FOR SOME Fe^{III} AND Co^{III} OXIDATIONS OF THE VANADIUM ION

Vanadium ion	Reaction with Fe^{III}	Reaction with Co^{III}
V^{II}	$>10^5$	>300
V^{III}	0·009	0·192
V^{IV}	~0·0025	0·260

The concentration of V^{IV} may be assumed equal to the concentration of reacted V^{III}, i.e. the concentration of V^{V} is small and negligible since (3) is fast.

In Table 24, rate constants are compared for Co^{III} and Fe^{III} reactions with V^{II}, V^{III} and V^{IV}, respectively, in 1 N perchloric acid solutions. The rate constant for the reaction of V^{IV} with Fe^{III} has been calculated knowing that for the reverse reaction, and the overall equilibrium constant,

$$VO^{2+} + Fe^{3+} + H_2O \rightleftharpoons VO_2^+ + Fe^{2+} + 2H^+,$$

($K = 10^{-6}\ l^{-2}\ mole^2$ at 0°C). When V^{IV} and Fe^{III} are mixed, there is no detectable reaction, since the above equilibrium lies very much to the left and net conversion to V^{V} and Fe^{II} is negligible. The similarity of the relative rates for the Co^{III} and Fe^{III} reactions in Table 24 is of some interest since it suggests that reactions proceed by the same electron-transfer mechanism. If, in the Co^{III} reactions, there is an initial electronic change (as suggested previously), further details of the reaction between V^{II} and Co^{III} could be of interest, since the reaction may be sufficiently fast for the conversion of spin-paired Co^{III} to spin-free Co^{III} to be rate determining.

A feature of reactions in which one of the reactants becomes more hydrolysed during the course of the reaction, e.g. when V^{3+} is oxidized to VO^{++}, are the entropies of activation, which are more positive than might otherwise be expected. Thus, for the reaction of V^{3+} with Co^{3+}, $\Delta S^{\ddagger} = 5 \pm 5$ e.u., in the reaction of V^{3+} with Fe^{3+}, $\Delta S^{\ddagger} = -15 \pm 13$ e.u., and in the reaction of V^{3+} with Cu^{2+}, $\Delta S^{\ddagger} = 5 \pm 5$ e.u. Such values can be accounted for by considering partial hydrolysis, i.e. loss of H^+ ions, from a coordinated water within the transition complex, the latter assisting in the conversion of V^{3+} to VO^{2+}.

The reaction between vanadium(III) and neptunium(V)

The reaction of V^{III} with Np^V, overall equation

$$V^{III} + Np^{V} \xrightarrow{k_1} V^{IV} + Np^{IV}, \tag{1}$$

also proceeds by an indirect route as well as that indicated by the stoicheiometric equation.[22] In this case, however, Np^{III} and not V^V is formed as an intermediate. Thus the additional steps are

$$V^{III} + Np^{IV} \overset{K_2}{\rightleftharpoons} V^{IV} + Np^{III} \tag{2}$$

$$Np^{III} + Np^{V} \xrightarrow{k_3} 2Np^{IV}. \tag{3}$$

Although the equilibrium in (2) is unfavourable ($K_2 = 6 \times 10^{-4}$ at 0°C), k_3 is sufficiently fast for this second route to be effective. Assuming stationary-state kinetics for Np^{III}, the rate equation

$$\frac{-d[V^{III}]}{dt} = [V^{III}][Np^{V}]\left(k_1 + \frac{k_2 k_3 [Np^{IV}]}{k_{-2}[V^{IV}]}\right)$$

22. E. H. Appelman and J. C. Sullivan, *J. Phys. Chem.* 66, 442 (1962).

can be derived. At 25°C and $\mu = 1{\cdot}0$, $k_1 = 0{\cdot}3$ l mole^{-1} sec^{-1} and $k_2k_3/k_{-2} = 0{\cdot}16$ l mole^{-1} sec^{-1}. The equilibrium constant K_2 and rate constant k_3 can be determined by separate experiment and the product gives satisfactory agreement with k_2k_3/k_{-2}.

Single-stage two-electron changes

For reactions in which both reactants change their oxidation number by two units, simultaneous or near-simultaneous two-electron transfers are possible. As in the isotopic exchange between thallium(I) and thallium(III), these are difficult to prove by straightforward kinetic methods. For example, the reaction of uranium(IV) with thallium(III),

$$U^{IV} + Tl^{III} \rightarrow U^{VI} + Tl^{I},$$

conforms to a rate equation

$$-\frac{d[Tl^{III}]}{dt} = k[U^{IV}][Tl^{III}],$$

but no information is obtained regarding the possible intermediate formation of U^{V} and Tl^{II}.[23] Thus, a first step

$$U^{IV} + Tl^{III} \rightarrow U^{V} + Tl^{II},$$

followed by the fast reaction

$$U^{V} + Tl^{II} \xrightarrow{\text{fast}} U^{VI} + Tl^{I},$$

would give second-order kinetics, as in the direct two-electron process.

One-equivalent changes would not necessarily provide an easier route as far as stepwise hydrolysis of uranium is concerned. Thus, increase in hydrolysis would occur in the first stage of a two-stage process, i.e. the $U^{IV} \rightarrow U^{V}$ conversion, since the uranium-(IV), (V) and (VI) ions are U^{4+}, UO_2^{+} and UO_2^{2+},

---

23. A. C. Harkness and J. Halpern, *J. Am. Chem. Soc.* **81**, 3526 (1959).

respectively. Contrast the reaction of vanadium(III) with thallium(III) (see below), in which the relevant vanadium ions are V^{3+}, VO^{2+}, and VO_2^+, and two one-equivalent steps might provide a more favourable path for reaction.

Occasionally, there is more positive evidence for a two-equivalent change, as in the reaction of chromium(II) and thallium(III),[24] which is thought to proceed by a reaction sequence

$$Cr^{II} + Tl^{III} \rightarrow Cr^{IV} + Tl^{I} \quad (1)$$

$$Cr^{II} + Cr^{IV} \rightarrow (Cr^{III})_2. \quad (2)$$

If mononuclear Cr^{III} were formed as an intermediate in (1), then a much greater proportion would be expected in the final product, since Cr^{III} is not easily oxidized to Cr^{IV}. It can also be reasoned that since Cr^{II} has an electronic configuration $t_{2g}^3 e_g^1$, with each electron in a different orbital, successive one-electron transfers within the lifetime of a single transition complex might seem more likely than simultaneous transfer of two electrons.

One way of establishing whether a highly reactive oxidation state is formed as an intermediate is to add a substrate, which is inert to reactants and products taken separately, but will react with the intermediate. Complexes of cobalt(III), e.g. $Co(C_2O_4)_3^{3-}$, have been used in this way to identify Sn^{III} in some Sn^{II} reactions in hydrochloric acid solutions. In the reactions with Cr^{VI}, Mn^{VII}, Ce^{IV} and Fe^{III}, Sn^{III} can be detected, but with Tl^{III}, Hg^{II}, H_2O_2, I_2 and Br_2, there is no evidence for the formation of intermediate Sn^{III}.[25]

The reaction between vanadium(III) and thallium(III)

At 5°C the reaction of V^{III} with Tl^{III} is too fast to be followed, using conventional techniques. Information regarding the mechanism can, however, be obtained by spectrophotometric

---

24. M. ARDON and R. A. PLANE, *J. Am. Chem. Soc.* **81**, 3197 (1959).
25. W. C. E. HIGGINSON and E. A. M. WETTON, *J. Chem. Soc.* (1965).

measurements on the final reacted solution (ref. 15, page 170). There are two possible mechanisms to consider,

(a) The one-stage two-equivalent change

$$V^{III} + Tl^{III} \rightarrow V^{V} + Tl^{I},$$

followed by

$$V^{III} + V^{V} \rightarrow 2V^{IV}$$

and

(b) a series of one-equivalent changes with the formation of unstable Tl^{II},

$$V^{III} + Tl^{III} \rightarrow V^{IV} + Tl^{II}$$

$$V^{IV} + Tl^{II} \rightarrow V^{V} + Tl^{I}$$

$$V^{III} + Tl^{II} \rightarrow V^{IV} + Tl^{I}$$

$$V^{III} + V^{V} \rightarrow 2V^{IV},$$

The reaction of V^{IV} with Tl^{III} is slow at 80°C and can be ignored in the above mechanism. Both (a) and (b) might give small concentrations of V^{V} in the final reacted solution, but only if (b) holds, will the amount of V^{V} be increased by adding V^{IV} initially Formation of V^{V} is not observed in perchloric acid solutions, but in dilute sulphuric acid, and with an excess of Tl^{III}, upto 3% of the V^{III}, is converted to V^{V}. This amount can be increased to 10% for experiments in which V^{IV} is present initially, which is consistent with the one-electron changes as indicated in (b).

The reaction between tin(II) and vanadium(V)

Alternative reaction paths are also possible in the rapid reaction of Sn^{II} with V^{V},[26] and information regarding the mechanism has, likewise, been obtained by spectrophotometric measurements on final reacted solutions. In dilute hydrochloric acid with an excess of Sn^{II}, a mixture of V^{III} and V^{IV} is obtained. The

---

26. D. J. DRYE, W. C. E. HIGGINSON and P. KNOWLES, *J. Chem. Soc.* 1137 (1962).

dependence of the concentration of these products upon reactant and excess V^{IV} concentrations, shows that the dominant initial step involves a two-equivalent change with the formation of V^{III},

$$Sn^{II} + V^{V} \rightarrow Sn^{IV} + V^{III}.$$

Other possible paths involving one-equivalent changes and the formation of unstable Sn^{III} are thought to make some contribution, however, and it has been suggested that the association complex $Sn^{II} . V^{V}$ may persist sufficiently long for the reaction

$$Sn^{II} . V^{V} + V^{V} \rightarrow 2V^{IV} + Sn^{IV},$$

to occur to a limited extent. The primary reaction processes can be represented by equations

$$Sn^{II} + V^{V} \rightarrow Sn^{II} . V^{V} \begin{cases} \nearrow Sn^{IV} + V^{III} & \text{(A)} \\ \rightarrow Sn^{III} + V^{IV} & \text{(B)} \\ \xrightarrow{V^{V}} Sn^{IV} + 2V^{IV} & \text{(C)} \end{cases}$$

where the relative rates for the different paths are A > B > C.

Formation of binuclear intermediates

Kinetically inert (but not necessarily thermodynamically stable) binuclear intermediates have been identified in a number of redox reactions. Typical cases involve Cr^{II} or Co^{II} as one of the reactants, the non-labile Cr^{III} and Co^{III} ions so formed capturing in their inner-coordination sphere a non-labile group attached to the other reactant.

In the Cr^{II} reduction of U^{VI},[27] the reactants disappear much faster than the final Cr^{III} and U^{IV} products are formed, and an inert intermediate,

$$Cr^{II} + U^{VI} \rightarrow Cr^{III}U^{V},$$

having a structure

$$[(H_2O)_5Cr^{III}—O—UV—O]^{4+},$$

27. G. Gordon, *Inorg. Chem.* **2**, 1277 (1963).

is believed to be formed. This then undergoes a much slower reaction with a second Cr^{II} ion, and a transition complex,

$$[(H_2O)_5Cr—O—U—O—Cr(H_2O)_5]^{6+},$$

is formed which decomposes to give two chromium(III) ions and one uranium(IV) as products. Experiments in which the UO_2^{2+} was labelled with oxygen-18 have indicated oxygen transfer from UO_2^{2+} (reactant) to $Cr(H_2O)_6^{3+}$ (product).

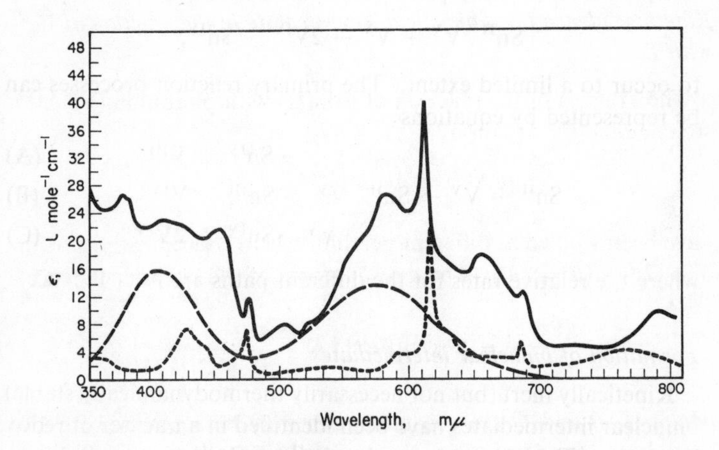

FIG. 27. Absorption spectra of Cr^{III} . Np^{V} (——), Cr^{III} (- - -) and Np^{V} (. . .). [Reproduced, with permission, from J. C. SULLIVAN, *J. Am. Chem. Soc.* **84**, 4257 (1962).]

A stable binuclear complex,

$$[(H_2O)_5Cr^{III}—O—Np^{V}—O]^{4+},$$

is also formed in the Cr^{II}–Np^{VI} reaction, and can be separated using ion-exchange resins. The Np/Cr ratio has been confirmed by analysis, and in Fig. 27 the spectrum of the complex is compared with spectra for Cr^{III} and Np^{V}.

In a further instance, a stable binuclear complex is formed when the pentacyano complex $Co^{II}(CN)_5^{3-}$ is oxidized by the $Fe^{III}(CN)_6^{3-}$ complex. A cyanide group on the oxidant is held in the inner-coordination sphere of the newly formed cobalt(III), with the formation of

$$[(CN)_5Fe^{II}CNCo^{III}(CN)_5]^{6-}.$$

The latter has been isolated.[28]

Kinetic evidence for a binuclear intermediate in the reaction of Fe^{II} with Pu^{VI}

The kinetics of the reaction of iron(II) with plutonium(VI),[29] stoicheiometric equation

$$Pu^{VI} + Fe^{II} \rightarrow Pu^{V} + Fe^{III},$$

have been studied at different perchloric acid concentrations from 0·05–2·0 N ($\mu = 2{\cdot}0$). The unusual form of the acid dependence,

$$\frac{-d[Pu^{VI}]}{dt} = [Pu^{VI}][Fe^{II}]\left(A + \frac{1}{B + C[H^+]}\right),$$

where A, B and C are constants, can only be accounted for by considering the formation of a metastable binuclear intermediate $Pu^{V}\,.\,Fe^{III}$, which is of the same composition as the transition complex. Thus, electron transfer can take place in the normal way,

$$PuO_2^{2+} + Fe^{2+} \rightarrow PuO_2^{+} + Fe^{3+}_{V} \qquad (1)$$

but the reactants can also interact with the formation of a binuclear intermediate,

$$PuO_2^{2+} + Fe^{2+} \rightarrow PuO_2\,.\,Fe^{4+}, \qquad (2)$$

28. A. Haim and W. K. Wilmarth, *J. Am. Chem. Soc.* **83**, 509 (1961).
29. T. W. Newton and F. B. Baker, *J. Phys. Chem.* **67**, 1425 (1963).

where $PuO_2 . Fe^{4+}$ and hydrolysed $PuO_2 . Fe^{4+}$ subsequently decompose to give products,

$$PuO_2 . Fe^{4+} \rightarrow PuO_2^+ + Fe^{3+}, \qquad (3)$$

$$PuO_2 . FeOH^{3+} \rightarrow PuO_2^+ + FeOH^{2+}. \qquad (4)$$

By assuming stationary-state kinetics for the metastable intermediate, eqns. (1)–(4) give a rate law in agreement with experiment.

The reactions of some dicobalt complexes

A whole range of dicobalt ammine complexes are known in which NH_2, OH, SO_4, NH, NO_2 and O_2 can function as bridging groups. Those with an O_2^- bridge are somewhat unusual in that oxidized (e.g. $(NH_3)_5 Co . O_2 . Co(NH_3)_5^{5+}$) and reduced (e.g. $(NH_3)_5 Co . O_2 . Co (NH_3)_5^{4+}$) forms can be prepared. The bridges are believed to be essentially O_2^- and O_2^{2-} respectively.

The kinetics of the 1 : 1 electron transfer reaction between the superoxo complex $(NH_3)_5Co . O_2 . Co(NH_3)_5^{5+}$ and Fe^{II}.

$$(NH_3)_5 Co . O_2 . Co(NH_3)_5^{5+} + Fe^{II} \rightarrow (NH_3)_5 Co . O_2 . Co(NH_3)_5^{4+} + Fe^{III} \qquad (1)$$

have been studied in perchloric acid solutions at $\mu = 2{\cdot}0$ M.[30] The 4 + peroxo complex which is formed in (1) is extremely unstable in acid solutions, and readily decomposes to give mononuclear Co^{II} ions (which are labile)

$$(NH_3)_5Co . O_2 . Co(NH_3)_5^{4+} + 10H^+ \xrightarrow{\text{fast}} 2Co^{2+} + O_2 + 10NH_4^+. \qquad (2)$$

Second-order rate constants k_{obs} as in equation (1) have a hydrogen-ion dependence of the general form

$$k_{obs} = a[H^+]^{-1} + b + c\,[H^+]$$

30. A. G. Sykes, *Trans. Faraday Soc.* **58,** 543 (1962), and **59,** 1325, 1334 (1963).

More recent work on (1) has shown that *a* and *c* make negligible contributions when μ is adjusted with $LiClO_4$. At 9°C the rate constant *b* for the hydrogen-ion independent path is 0·028 $mole^{-1} sec^{-1}$ and activation parameters are $\Delta H_b^\ddagger = 6{\cdot}9$ kcal $mole^{-1}$ and $\Delta S_b^\ddagger = -41{\cdot}5$ e.u. For the corresponding reaction of the double bridged complex

$$(NH_3)_4Co \,.\, \mu(NH_2, O_2) \,.\, Co(NH_3)_4^{4+}$$

a similar hydrogen-ion dependence is observed, but rate constants are considerably bigger (*b* is 25·2 $mole^{-1} sec^{-1}$ at 20°C). In these reactions the peroxo group is believed to be the centre at which electron-transfer occurs, steric effects being important in determining the difference in rates. Thus, the dibridged complex, in which there is more ready access to the peroxo group, reacts considerably faster than the mono bridged complex, in which close approach of the Fe^{2+} to the peroxo group is probably hindered by coordinated ammonia groups. The effect which different anions X^- have on the reaction of

$$(NH_3)_5Co \,.\, O_2 \,.\, Co(NH_3)_5^{5+}$$

with Fe^{II} has also been studied. The X^- dependent reaction steps can be expressed

$$k_{cat} = k'[X^-] + k''[H^+][X^-]$$

where values for k' (i.e. data at low H^+ concentrations) are shown in Table 25.

TABLE 25

A COMPARISON OF THE EFFECTIVENESS OF DIFFERENT ANIONS X^- ON THE REACTION OF Fe^{II} WITH $(NH_3)_5Co \,.\, O_2Co(NH_3)_5^{5+}$ AT 25°C, $\mu = 2{\cdot}0$ M

X^-	k' l^2 $mole^{-2} sec^{-1}$
F^-	6900
Cl^-	392
SO_4^{2-}	193
Br^-	84
NO_3^-	<·3

In other reactions of the single bridged 5+ complex with V^{IV}, Sn^{II}, I^- and $S_2O_3^{2-}$ there is evidence for 1 : 1 stoicheiometries as in the reaction with Fe^{II}. With SO_3^{2-} and NO_2^- however 1 : 2 stoicheiometries are observed, no oxygen is evolved and a Co^{III} complex is formed e.g.

$$(NH_3)_5Co \,.\, O_2 \,.\, Co(NH_3)_5^{5+} + 2SO_3^{2-} \xrightarrow{5H^+} Co(NH_3)_5SO_4^- + Co^{II} + SO_4^{2-} + 5NH_4^+.$$

While the 1 : 1 reactions are thought to proceed by an electron-transfer mechanism, the 1 : 2 reactions probably involve transfer of the peroxo oxygen atoms to the oxyanions.

CHAPTER 9

SOME REACTIONS OF IONS AND MOLECULES OF NON-METALLIC ELEMENTS IN AQUEOUS SOLUTIONS

THE kinetics and mechanisms of reactions involving ions and molecules of non-metallic elements are described in this chapter. In addition to the reactions of hydrogen peroxide which have been of considerable interest to chemists for some time, reactions of hydrazine, hydroxylamine, iodide, sulphite, persulphate and oxy-halogen ions are considered. A feature of these reactions are the bond-breaking processes which often result in the formation of free radical intermediates, such intermediates being analogous to the unstable oxidation states in the reactions of metal ions. In the final section, some induced reactions involving iodide and arsenite ions are considered.

Some reactions of iodide ions

The reaction of iodide ions with hydrogen peroxide,[1] stoicheiometric equation

$$2I^- + H_2O_2 + 2H^+ \rightarrow 2H_2O + I_2,$$

obeys a rate equation of the form

$$d[I^-]/dt = k_1[I^-][H_2O_2] + k_2[H^+][I^-][H_2O_2].$$

1. H. A. LIEBHAFSKY and A. MOHAMMAD, *J. Am. Chem. Soc.* **55**, 3977 (1933).

This suggests that there are two parallel rate determining steps,

$$I^- + H_2O_2 \xrightarrow{k_1} H_2O + IO^- \tag{1}$$

and

$$I^- + H^+ + H_2O_2 \xrightarrow{k_2} H_2O + HOI, \tag{2}$$

where, in (2), the reaction could be between HI and H_2O_2 or, alternatively, I^- and $H_3O_2^+$. The latter path is by no means unreasonable since hydrogen iodide is a strong acid,

$$HI \rightleftharpoons H^+ + I^-, \ (K \sim 10^7)$$

and the concentration of un-ionized hydrogen iodide is extremely small. The extent of the interaction of hydrogen peroxide with protons is not yet known, but since there are electron " lone-pairs " on each of the oxygen atoms, some degree of protonation is clearly possible. Subsequent reactions

$$I^- + HOI \rightarrow I_2 + OH^-$$

and

$$H^+ + OH^- \rightarrow H_2O,$$

are fast and non-rate determining.

The reaction between iodide ions and ferric,

$$2I^- + 2Fe^{3+} \rightarrow I_2 + 2Fe^{2+}$$

has been studied in perchloric acid solutions.[2] The main features of the concentration dependence are included in the rate equation

$$\frac{-d[Fe^{3+}]}{dt} = \frac{a[I^-]^2[Fe^{3+}]}{b[Fe^{2+}]/[Fe^{3+}] + 1},$$

2. K. W. Sykes, *J. Chem. Soc.* 124 (1952).

which is consistent with a mechanism

$$I^- + Fe^{3+} \underset{k_{-1}}{\overset{k_1}{\rightleftharpoons}} FeI^{2+}$$

$$I^- + FeI^{2+} \underset{k_{-2}}{\overset{k_2}{\rightleftharpoons}} Fe^{2+} + I_2^-$$

$$I_2^- + Fe^{3+} \xrightarrow{k_3} I_2 + Fe^{2+}.$$

Making two stationary-state assumptions for $[I_2^-]$ and $[FeI^{2+}]$, and solving for $[I_2^-]$, the overall rate

$$-d[Fe^{3+}]/dt = 2k_2[I_2^-][Fe^{3+}],$$

may be expressed

$$\frac{-d[Fe^{3+}]}{dt} = \frac{2k_1k_2k_3[I^-]^2[Fe^{3+}]}{k_{-1}k_{-2}[Fe^{2+}]/[Fe^{3+}]+k_{-1}k_3+k_2k_{-2}[Fe^{2+}][I^-]/[Fe^{3+}]+k_3k_2[I^-]}$$

Assuming the last two terms on the lower line to be small, this is of the same form as the experimental rate equation.

Other simple anions X^- have an inhibitory effect on the reaction, since they complex with the ferric, in this way limiting the formation of FeI^{2+}. Assuming the reaction of iodide with FeX^{2+} to be negligible, stability constants for the formation of FeX^{2+} can be determined. These show good agreement with existing values.

The decomposition of hydrogen peroxide

The reaction of ferrous and ferric ions with hydrogen peroxide has been extensively studied.[3] With equimolar amounts of Fe^{II} and hydrogen peroxide and with the hydrogen-ion concentration $<0{\cdot}01$ N, there is a nearly quantitative reaction

$$2Fe^{II} + H_2O_2 + 2H^+ \rightarrow 2Fe^{III} + 2H_2O,$$

3. W. G. Barb, J. H. Baxendale, P. George and K. R. Hargrave, *Trans. Faraday Soc.* 47, 462 (1951).

and the rate law[4] is of the form

$$-d[\mathrm{Fe^{II}}]/dt = k[\mathrm{Fe^{II}}][\mathrm{H_2O_2}].$$

Under these conditions the mechanism is essentially

$$\mathrm{Fe^{II} + H_2O_2 \rightarrow FeOH^{2+} + OH}$$

$$\mathrm{Fe^{II} + OH \xrightarrow{fast} FeOH^{2+}},$$

the subsequent reaction of $FeOH^{2+}$ with H^+ being rapid. Evidence supporting the formation of OH radicals has been obtained using monomer substrates. Thus, the radicals will initiate the polymerization of methyl methacrylate, the latter competing with the ferrous for reaction with hydroxyl radicals. With increasing amounts of the monomer, the ratio of Fe^{II} to hydrogen peroxide consumed, decreases from just less than 2 : 1 in the absence of monomer to 1 : 1 in the presence of a large excess of monomer.

Over long periods, there is a slow evolution of oxygen from the Fe^{II}–H_2O_2 system, and stoicheiometric measurements indicate a catalytic decomposition of hydrogen peroxide. This effect can be increased by adding Fe^{III}, by increasing the concentration of hydrogen peroxide, and by increasing the hydrogen-ion concentration. The decomposition is thought to involve the ferric oxidation of hydrogen peroxide, the HO_2 radical which is first formed being oxidized by a second ferric ion to oxygen. This is supported by oxygen-18 tracer experiments which have shown that both oxygen atoms in the oxygen produced came from the same hydrogen peroxide molecule.[5] At the same time, the ferrous reduces hydrogen peroxide to water, so that the net effect of the Fe^{II}–Fe^{III} couple is to catalyse the disproportionation of hydrogen peroxide,

$$\mathrm{2H_2O_2 \rightarrow 2H_2O + O_2}.$$

4. N. Uri, *Chem. Rev.* **50**, 375 (1951).
5. C. A. Bunton and D. R. Llewellyn, *Research* (*London*) **5**, 142 (1952).

A mechanism which seems to fit all the experimental data is as follows

Initiation steps $Fe^{II} + H_2O_2 \rightleftharpoons Fe^{III} + OH^- + OH$

$$Fe^{III} + H_2O_2 \rightleftharpoons Fe^{II} + HO_2 + H^+$$

Propagation steps $OH + H_2O_2 \rightarrow H_2O + HO_2$

$$Fe^{III} + HO_2 \rightarrow Fe^{II} + O_2 + H^+$$

$$Fe^{II} + H_2O_2 \rightarrow Fe^{III} + OH^- + OH$$

Termination $Fe^{II} + OH \rightarrow Fe^{III} + OH^-$.

Depending on the pH, further reactions involving O_2H^- and O_2^- can be included, where the latter are formed by ionization of H_2O_2 and HO_2. Similar reaction paths are possible with other elements having different oxidation numbers in place of the ferrous and ferric.

The reduction of hydroxylamine

In acid solutions the reduction of hydroxylamine with Ti^{III} is known to be stoicheiometric,

$$2Ti^{III} + NH_2OH + 3H^+ \rightarrow 2Ti^{IV} + NH_4^+ + H_2O,$$

and is used in the quantitative determination of hydroxylamine. The reaction sequence is not immediately apparent, however, since NH_2 radicals or OH radicals or both may be involved. Possible mechanisms are

$$Ti^{III} + NH_2OH \rightarrow Ti^{IV} + NH_2 + OH^-$$

$$Ti^{III} + NH_2 \xrightarrow{H^+} Ti^{IV} + NH_3, \qquad (1)$$

and

$$Ti^{III} + NH_2OH \rightarrow Ti^{IV} + NH_2^- + OH$$

$$Ti^{III} + OH \xrightarrow{H^+} Ti^{IV} + H_2O, \qquad (2)$$

where NH_2OH is present as NH_3OH^+, and the Ti^{IV} ions are extensively hydrolysed. To distinguish between these possibilities, a monomer substrate can be added,[6] when the monomer and Ti^{III} compete for the radicals present. For each monomer–radical reaction, a polymerization is initiated, and the radical is, in effect, removed from the system. At completion, the free ammonia can be determined by the Kjeldahl method. If NH_2 radicals are formed, then the amount of ammonia produced will decrease as the ratio of monomer to Ti^{III} is increased. But if, on the other hand, OH radicals only are formed, then the amount of ammonia will be unaffected by the presence of monomer. For a series of runs with varying reactant and monomer concentrations, the relationship

$$[Ti^{III}]_{consumed} = [NH_2OH]_{consumed} + [NH_3]_{formed}$$

was found to hold, and it can be concluded that the reaction proceeds exclusively by path (1) with the intermediate formation of NH_2 radicals.

The oxidation of hydrazine

When aqueous solutions of hydrazine are oxidized, possible reaction products are nitrogen, ammonia and, at higher acid concentrations, hydrazoic acid.[7] The reactions are of particular interest since the products allow a distinction to be made between one-equivalent (e.g. Ce^{IV}, Mn^{III}, Fe^{III}, Co^{III}), and two-equivalent (e.g. Tl^{III}, Cl_2, Br_2, I_2, BrO_3^-, IO_3^-) oxidizing agents. In 0·01–0·10 N sulphuric acid, and at 20°C, all these reactants give nitrogen; with one-equivalent reagents, ammonia is also formed. The relevant reaction paths are thought to be

$$N_2H_4 \xrightarrow{-e} N_2H_3 \rightarrow \tfrac{1}{2}N_4H_6 \rightarrow \tfrac{1}{2}N_2 + NH_3 \quad (1)$$

$$N_2H_4 \xrightarrow{-2e} N_2H_2 \xrightarrow{-2e} N_2 \quad (2)$$

6. P. Davis, M. G. Evans and W. C. E. Higginson, *J. Chem. Soc.* 2563 (1951).
7. W. C. E. Higginson, *The Oxidation of Hydrazine in Aqueous Solution*, Special Publ. no. 10, p. 95, The Chemical Society, London, 1957.

where one-equivalent reagents react mostly by (1), and two-equivalent reagents exclusively by (2).

Evidence in support of the above mechanism has been obtained using nitrogen-15 enriched hydrazine. Thus the nitrogen formed in (2), overall equation

$$N_2H_4 - 4e \rightarrow N_2 + 4H^+,$$

has the same isotopic distribution as the original hydrazine, and both nitrogen atoms must, therefore, orginate from the same molecule. With one-equivalent reagents, on the other hand, when the overall equation is

$$N_2H_4 - e \rightarrow \tfrac{1}{2}N_2 + NH_3 + H^+,$$

half the nitrogen contains both atoms from one hydrazine, but, with the other half, the two nitrogen atoms come from different molecules. This is consistent with the intermediate formation of N_4H_6 since the nitrogen chain in the latter can break up in two ways. Thus in the first step an N_3 compound is formed,

$$N^a{-}N^a{-}N^b{-}N^b \quad \rightarrow \quad N^a{-}N^a{-}N^b + N^b,$$

where the latter has resonance forms

$$N^a = N^a{-}N^b \quad \rightleftharpoons \quad N^a{-}N^a = N^b,$$

and further decomposition can give

$$N^a = N^a{-}N^b \quad \rightarrow \quad N^a \equiv N^a + N^b$$

and

$$N^a{-}N^a = N^b \quad \rightarrow \quad N^a + N^a \equiv Nb.$$

In a detailed study of the reaction of hydrazine with Fe^{III}, the kinetics are in agreement with the basic reaction sequence

$$N_2H_4 + Fe^{III} \quad \rightleftharpoons \quad N_2H_3 + Fe^{II} + H^+$$

$$2N_2H_3 \quad \rightarrow \quad N_4H_6$$

$$N_4H_6 \quad \xrightarrow{\text{fast}} \quad 2NH_3 + N_2.$$

In addition, there are further contributions from the reactions

$$N_2H_3 + Fe^{III} \rightarrow N_2H_2 + Fe^{II} + H^+$$

$$N_2H_2 + 2Fe^{III} \rightarrow N_2 + 2Fe^{II} + 2H^+,$$

since one-equivalent oxidizing agents can also react by reaction path (2), which is applicable to two-equivalent oxidizing agents.

In 1–10 N sulphuric acid and at ~80°C hydrazoic acid, HN_3, is formed in the reaction with two-equivalent reagents. To account for this, path (2) may be amended in the following way to include the formation and subsequent decomposition of the N_4H_4 species. Thus

$$N_2H_4 \xrightarrow{-2e} N_2H_2 \rightarrow N_4H_4 \rightarrow NH_3 + HN_3.$$

Little or no hydrazoic acid is obtained with one-equivalent oxidizing agents.

The oxidation of sulphite solutions

A distinction between one- and two-equivalent oxidizing agents is also possible in the oxidation of sulphite solutions,[8]

$$SO_3^{2-} \xrightarrow{-1e} SO_3^- \rightarrow \tfrac{1}{2}S_2O_6^{2-} \qquad (1)$$

$$SO_3^{2-} \xrightarrow{-2e} SO_3 \quad \text{(i.e. } SO_4^{2-}\text{).} \qquad (2)$$

Thus, one-equivalent reagents can react by both (1) and (2), giving dithionate and sulphate (stoicheiometry between 1·0 and 2·0), and two-equivalent reagents can react by (2) giving sulphate only (stoicheiometry 2·0). The stoicheiometries are readily determined using standard volumetric procedures, and an excess of unreacted sulphite can, for example, be determined by titrating with iodine. The results shown in Table 26 are at pH 0·5, when the sulphite is present largely as H_2SO_3 and SO_2. A number of

8. W. C. E. Higginson and J. Marshall, *J. Chem. Soc.* 447 (1957).

the reactions have been studied at higher pHs, when the predominant reducing species are HSO_3^- (at pH 5), and SO_3^{2-} (at pH 9). Under these conditions the stoicheiometries are similar to those obtained at the lower pH.

TABLE 26

STOICHEIOMETRIES FOR THE REACTION OF ONE- AND TWO-EQUIVALENT OXIDIZING AGENTS WITH H_2SO_3 AT pH 0·5

One-equivalent reagents	Stoicheiometry	Two-equivalent reagents	Stoicheiometry
Ce^{IV}	1·27–1·44	I_2	2·0
Co^{III}	1·04–1·37	Br_2	2·0
Fe^{III}	~1·2	Cl_2	2·0
*V^V	1·57	IO_3^-	2·0
*Mn^{VII}	1·55–1·80	H_2O_2	2·0
*Cr^{VI}	1·84–1·95	Tl^{III}	2·0

* One- and two-equivalent reactions are possible and both may contribute.

A kinetic study has been made of the reaction of H_2SO_3 with Fe^{III}, but the results do not allow a unique interpretation to be made. In the presence of Cu^{II} as catalyst, results are consistent with a mechanism

$$H_2SO_3 + Fe^{III} \overset{-H^+}{\rightleftharpoons} HSO_3 + Fe^{II} \quad (1)$$

$$2HSO_3 \rightarrow H_2S_2O_6 \quad (2)$$

$$HSO_3 + Fe^{III} \overset{-H^+}{\rightarrow} SO_3(H_2SO_4) + Fe^{II} \quad (3)$$

$$HSO_3 + Cu^{II} \overset{-H^+}{\rightarrow} SO_3(H_2SO_4) + Cu^{I} \quad (4)$$

followed by

$$Cu^{I} + Fe^{III} \overset{\text{fast}}{\rightarrow} Cu^{II} + Fe^{II}. \quad (5)$$

The catalysis is dependent on reaction (4) being faster than (3). When Fe^{II} is added initially, the reaction is retarded, thus supporting the inclusion of the back reaction in (1).

Some reactions involving oxygen-atom transfer

In some further reactions of sulphite ions with oxyhalogen compounds ClO_3^-, BrO_3^-, ClO_2, ClO_2^-, Cl_2O and ClO^-, oxygen atom transfer has been demonstrated using oxygen-18 as a tracer element.[9] The reactions are believed to take place stepwise, e.g.

$$ClO_3^- \rightarrow ClO_2^- \rightarrow ClO^- \rightarrow Cl^-,$$

and in acid solutions the final products are, in each case, sulphate and halide ions. That complete oxygen-18 transfer is not observed has been attributed to an alternative path in the reaction of ClO^- with SO_3^{2-}. Thus, in addition to the direct transfer,

$$SO_3^{2-} + Cl^{18}O^- \rightarrow SO_3{}^{18}O^{2-} + Cl^-,$$

oxidation can proceed by the alternative route

$$SO_3^{2-} + Cl^{18}OH \rightarrow ClSO_3^- + {}^{18}OH^-,$$

$$ClSO_3^- + H_2O \rightarrow SO_4^{2-} + HCl + H^+,$$

in which there is no oxygen-atom transfer from the hypochlorite to the sulphite.

Similar tracer studies have been possible for the oxidation of sulphite with hydrogen peroxide in acid solutions pH 1–5. The stoicheiometric equation is

$$H_2SO_3 + H_2O_2 \rightarrow H_2SO_4 + H_2O,$$

and, using labelled hydrogen peroxide, it can be shown that both oxygen atoms are transferred to the sulphite. In order to explain

9. J. HALPERIN and H. TAUBE, *J. Am. Chem. Soc.* **74**, 380 (1952).

this somewhat surprising result, intermediate formation of persulphurous acid has been suggested,

$$H{-}O{-}\overset{\overset{\displaystyle O}{|}}{S}{-}{}^{18}O{-}{}^{18}O{-}H,$$

which then rearranges to give doubly labelled sulphuric acid,

$$H{-}O{-}\overset{\overset{\displaystyle O}{|}}{\underset{\underset{\displaystyle {}^{18}O}{|}}{S}}{-}{}^{18}O{-}H.$$

Such a reaction sequence is similar to that observed for the reaction of nitrite ions with hydrogen peroxide, when pernitrous acid is known to be formed as an intermediate.

Further experiments using oxygen-18 have shown that, in the oxidation of sulphite by nitrite, the oxygen appearing on the sulphite is derived from the solvent.

The reduction of peroxydisulphate ions[10]

Although peroxydisulphate is one of the most powerful oxidizing agents known,

$$S_2O_8^{2-} + 2e \rightarrow 2SO_4^{2-}, \quad E^0 = 2{\cdot}01 \text{ V},$$

its reactions are generally slow unless a catalyst is present. In the oxidation of solutions of Ce^{III}, Cr^{III}, V^{IV} and N_2H_4 with Ag^I as catalyst,[11] the rate law is

$$-d[S_2O_8^{2-}]/dt = k[S_2O_8^{2-}][Ag^I],$$

and rate constants k differ only slightly as the reducing agent is varied. For any one experiment, $[Ag^I]$ may be assumed constant, so that pseudo-first order kinetics are observed. At 25°C and with the ionic strength around 0·5 M, k is *ca.* 0·005 l $mole^{-1}$ sec^{-1}.

10. D. A. House, *Kinetics and Mechanism of Oxidation by Peroxydisulphate, Chem. Rev.* **62,** 185 (1962).
11. W. M. Cone, *J. Am. Chem. Soc.* **67,** 78 (1945).

There are two possible mechanisms for the catalytic path, since Ag^{III} or Ag^{II} can be formed in the initial step. In the first of these possibilities the rate determining step

$$Ag^{I} + S_2O_8^{2-} \rightarrow Ag^{III} + 2SO_4^{2-}, \qquad (1)$$

is followed by the rapid equilibration of Ag^{III} with Ag^{I},

$$Ag^{I} + Ag^{III} \overset{\text{fast}}{\rightleftharpoons} 2Ag^{II}, \qquad (2)$$

so that either Ag^{II} or Ag^{III} might react subsequently with the reducing ion, e.g. with Ce^{III} reactions may be expressed

either $\quad Ce^{III} + Ag^{II} \rightarrow Ce^{IV} + Ag^{I},$

or $\quad Ce^{III} + Ag^{III} \rightarrow Ce^{IV} + Ag^{II}.$

In the second mechanism, Ag^{II} and the sulphate radical (for which there is evidence in other peroxydisulphate reactions) are formed in the rate determining step

$$Ag^{I} + S_2O_8^{2-} \rightarrow Ag^{II} + SO_4^{2-} + SO_4^{-}. \qquad (3)$$

Disproportionation of Ag^{II} to Ag^{I} and Ag^{III} is fairly rapid, so that again oxidation can proceed via Ag^{II} or Ag^{III}. What is interesting, but so far unexplained in these reactions, is the reason for the speed of the catalytic path with Ag^{I} and the relative slowness of uncatalysed reactions.

The reaction between iodine and peroxydisulphate ions,[12]

$$S_2O_8^{2-} + 2I^- \rightarrow 2SO_4^{-} + I_2,$$

is first order in both reactants. The mechanism is thought to be

$$S_2O_8^{2-} + I^- \rightarrow IS_2O_8^{3-}, \qquad (4)$$

$$IS_2O_8^{3-} + I^- \overset{\text{fast}}{\rightarrow} I_2 + 2SO_4^{2-}, \qquad (5)$$

where reaction (4), only, is rate determining. When reaction

12. E. S. Amis and J. E. Potts, *J. Am. Chem. Soc.* 63, 2883 (1941).

solutions are saturated with iodine, rate constants are reduced by about 25%, which has been attributed to the slower reaction of I_3^- compared to I^-. The full rate equation is of the form

$$-d[S_2O_8^{2-}]/dt = k_1[S_2O_8^{2-}][I^-] + k_2[S_2O_8^{2-}][I_3^-],$$

and at 25°C, with around 0·01 M concentrations of the reactants, $k_1 = 0{\cdot}0021$ and $k_2 = 0{\cdot}00107$ l mole^{-1} sec^{-1}.

A detailed study has been made of the reaction of hydrogen peroxide with peroxydisulphate at 30°C.[13] The stoicheiometry of the reaction does not deviate appreciably from that of the equation

$$S_2O_8^{2-} + H_2O_2 \rightarrow O_2 + 2HSO_4^-,$$

the empirical rate equation taking the form

$$\frac{-d[S_2O_8^{2-}]}{dt} = \frac{[S_2O_8^{2-}]}{\left[\dfrac{9{\cdot}5\times10^6}{[H_2O_2]} + \dfrac{7{\cdot}6\times10^7[S_2O_8^{2-}]}{[H_2O_2]} + 7{\cdot}9\times10^8 + 2{\cdot}9\times10^{10}[S_2O_8^{2-}]\right]^{1/2}}$$

In three different ranges of concentrations of reactants, this rate law approaches three different limiting forms,

(a) first-order in $[S_2O_8^{2-}]$, and half-order in $[H_2O_2]$,
(b) first-order in $[S_2O_8^{2-}]$,
(c) half-order in $[S_2O_8^{2-}]$.

The proposed mechanism involves a chain reaction with sulphate, hydroxyl, and HO_2 radicals carrying and terminating the chain. The chain initiation step is believed to be the decomposition reaction

$$S_2O_8^{2-} \rightarrow 2SO_4^-.$$

That sulphate radicals are formed in the thermal decomposition of

---

13. M. Tsao and W. K. Wilmarth, *Discussions Faraday Soc.* **29**, 137 (1960).

peroxydisulphate is evident from the fact that in polymerization studies, with peroxydisulphate labelled with sulphur-35 as an initiator, polymer fragments containing radioactive sulphate groups have been isolated. Subsequent chain carrying steps in the reaction with hydrogen peroxide are believed to be

$$SO_4^- + H_2O \rightarrow H^+ + SO_4^{2-} + OH$$

$$OH + H_2O_2 \rightarrow H_2O + HO_2$$

$$HO_2 + S_2O_8^{2-} \rightarrow O_2 + HSO_4^- + SO_4^-$$

$$HO_2 + H_2O_2 \rightarrow O_2 + H_2O + OH,$$

and termination reactions

$$HO_2 + OH \rightarrow O_2 + H_2O$$

$$SO_4^- + OH \rightarrow HSO_5^-$$

$$HO_2 + SO_4^- \rightarrow O_2 + HSO_4^-$$

$$SO_4^- + SO_4^- \rightarrow S_2O_8^{2-}.$$

The rate law describing the disappearance of $S_2O_8^{2-}$ may be derived by making the usual steady-state approximation and, since chain lengths are large, by neglecting initiation and termination steps. The resultant rate equation has the same general form as the experimental equation.

Some induced reactions

One of the earliest reaction systems to be recognized in this field was the aerial oxidation of arsenite in basic solutions in the presence of sulphite. The reaction between sulphite and oxygen alone is rapid under such conditions, but there is no corresponding reaction between arsenite and oxygen. In such instances, the sulphite is said to induce the reaction between the arsenite and oxygen, and may be termed an inductor; it is not by definition a catalyst, since it is itself undergoing reaction. In this instance, clearly, an intermediate formed during the conversion of sulphite to sulphate, overall reaction

$$2SO_3^{2-} + O_2 \rightarrow 2SO_4^{2-},$$

is able to effect the oxidation of arsenite to arsenate.

A system which, by its colour changes, provides a more ready demonstration of induction, is that involving chromium(VI), iodide and ferrous in solutions 0·001 N in acid. The oxidation of Cr^{VI} by iodide is normally slow, but in the presence of Fe^{II}, which reacts rapidly with Cr^{VI}, iodine is rapidly formed. Since in the Fe^{II} reduction of Cr^{VI} to Cr^{III}, unstable Cr^{V} and Cr^{IV} oxidation states are formed, one or both of these must be responsible for the oxidation of the iodide to iodine. The only other possible reaction, that between the Fe^{III} produced and iodide, can be shown to be slow under the conditions used. In the presence of a large excess of iodide, two equivalents of iodide are consumed for each equivalent of ferrous, and since the first step involving the latter is

$$Cr^{VI} + Fe^{II} \rightarrow Cr^{V} + Fe^{III},$$

the iodide must subsequently react with Cr^{V} to give Cr^{III}.[14] The overall reaction

$$Cr^{V} + 2I^{-} \rightarrow Cr^{III} + I_2,$$

may or may not involve intermediate formation of Cr^{IV}, since one- or two-equivalent changes may be effective. If the two-equivalent change

$$Cr^{V} + I^{-} \rightarrow Cr^{III} + I^{+}$$

is involved, this will be quickly followed by

$$I^{+} + I^{-} \rightarrow I_2.$$

Arsenite ions are also effective in inducing the reaction between Cr^{VI} and iodide, and, at high iodide concentrations, a ratio of

14. C. Wagner and W. Preiss, *Z. Anorg. Chem.* **168**, 265 (1928).

two equivalents of iodide to one of the inductor are likewise involved. It has been suggested that the mechanism of the reaction of As^{III} with Cr^{VI} (with no iodide present) is

$$As^{III} + Cr^{VI} \rightarrow As^{V} + Cr^{IV} \quad (1)$$

$$Cr^{IV} + Cr^{VI} \rightarrow 2Cr^{V} \quad (2)$$

$$Cr^{V} + As^{III} \rightarrow Cr^{III} + As^{V}. \quad (3)$$

In the presence of an excess of iodide reactions, (1) and (2) are presumably followed by the reaction of Cr^{V} and I^{-} as in the previous case.

The further instance in which arsenite ions induce a reaction between Cr^{VI} and Mn^{II} is of interest, since, with a large excess of Mn^{II}, a ratio of 0·5 equivalents of Mn^{II} to one of inductor are consumed.[15] This suggests that reaction (1) above is followed by

$$Cr^{IV} + Mn^{II} \rightarrow Cr^{III} + Mn^{III},$$

and that there is no reaction between Cr^{V} and Mn^{II}. Unless the hydrogen-ion concentration is very high, the Mn^{III} disproportionates,

$$2Mn^{III} \rightarrow Mn^{II} + Mn^{IV},$$

and MnO_2 is precipitated.

15. R. LANG and J. ZWERINA, *Z. Anorg. Chem.* **170**, 381 (1928).

CHAPTER 10

SOME FURTHER REACTIONS OF METAL IONS IN AQUEOUS SOLUTIONS

FURTHER reactions of metal ions with water, perchlorate ions and diatomic molecules are considered in this chapter. The reactions of strongly oxidizing ions (with water) and strongly reducing ions (with perchlorate ions and oxygen) are particularly relevant in considering the general stability of metal ions in solution. In reactions of Cr^{2+} with halogen molecules the products which are obtained indicate the closeness of approach of the reactants in the transition complex.

The reduction of silver(II) by water

Solutions of Ag^{II} perchlorate ($\sim 10^{-2}$ M) oxidize water at a fairly rapid rate, the half-life for reactions in 2·0 N perchloric acid being of the order of one to two hours.[1] The kinetics can be followed spectrophotometrically since Ag^{II} absorbs in the visible, and such measurements have shown that the reaction proceeds with the initial formation of Ag^{III}. Thus the rate equation is of the form

$$-d[Ag^{II}]/dt = k_{obs}[Ag^{II}]^2/[Ag^I],$$

Ag^I inhibiting the reaction. This is consistent with a reaction sequence

$$2Ag^{II} \underset{K}{\overset{\text{fast}}{\rightleftharpoons}} Ag^I + Ag^{III} \qquad (1)$$

$$Ag^{III} \xrightarrow{k_2} \text{Products}, \qquad (2)$$

1. J. B. KIRWIN, F. D. PEAT, P. J. PROLL and L. H. SUTCLIFFE, *J. Phys. Chem.* 67, 1617 (1963).

so that k_{obs} can be identified as k_2K. In the above, it can be seen that Ag^{I} inhibits the reaction by reducing the equilibrium concentration of Ag^{III}. From the inverse hydrogen-ion dependence of k_{obs}, it has been suggested that Ag^{III} is hydrolysed to AgO^+, which then reacts to give products,

$$AgO^+ \rightarrow Ag^+ + \tfrac{1}{2}O_2.$$

Clearly, the kinetics indicate negligible direct Ag^{II} oxidation of water.

The reduction of cobalt(III) by water

In solutions less than 0·1 N in perchloric acid, cobalt(III) perchlorate is reduced by water at a fairly rapid rate. At 25°C, for example, the half-life of a 2×10^{-3} M solution in 0·08 N acid is about twenty minutes, and this decreases to about eight minutes in 0·04 N acid. From kinetic experiments,(2) the rate has been shown to be of the order 3/2 in Co^{III} and inversely proportional to $[H^+]^2$. The Co^{III} dependence is difficult to account for, except by supposing that Co^{III} in aqueous solutions is present predominantly in a dimeric form. This is contrary to the accepted view, but there is, so far, no evidence which definitely rules out this possibility. The following rapid equilibrium reactions have been suggested,

$$Co^{3+} + H_2O \rightleftharpoons CoOH^{2+} + H^+$$

$$2CoOH^{2+} \xrightarrow{K_2} Co\text{—}O\text{—}Co^{4+} + H_2O$$

and

$$Co\text{–}O\text{–}Co^{4+} + 2H_2O \overset{K_3}{\rightleftharpoons} HOCo\text{–}O\text{–}CoOH^{2+} + 2H^+,$$

where K_2 is large and K_3 small. The rate determining step is then

$$CoOH^{2+} + HOCo\text{–}O\text{–}CoOH^{2+} \xrightarrow{k_4} 3Co^{2+} + 2OH^- + HO_2,$$

so that

$$-d[Co^{III}]/dt = k_4[CoOH^{2+}][HOCo\text{–}O\text{–}CoOH^{2+}].$$

2. J. H. Baxendale and C. F. Wells, *Trans. Faraday Soc.* **53**, 800 (1957).

Substituting for $[HOCo–O–CoOH^{2+}]$ and $[CoOH^{2+}]$

$$-\frac{d[Co^{III}]}{dt} = \frac{K_3k_4[Co–O–Co]^{3/2}}{K_2^{\frac{1}{2}}[H^+]^2}$$

and since Co^{III} is assumed to be present predominantly as $Co–O–Co^{4+}$, this can be written

$$-\frac{d[Co^{III}]}{dt} = \frac{k[Co^{III}]^{3/2}}{[H^+]^2},$$

which is in agreement with experiment.

The oxidation of metal ions with perchlorate ions

Although potentially a strong oxidizing agent perchlorate ions show little reactivity with metal ions. In only three instances is there appreciable reaction, these being with the fairly strong reducing ions vanadium(II) (t_{2g}^3), vanadium(III) (t_{2g}^2) and titanium(III) (t_{2g}^1). More recently[3] it has been shown that there is no reaction with the much stronger reducing ions europium(II) ($4f^7$) and chromium(II) ($t_{2g}^3e_g^1$). Thus with Eu^{II} there is no direct dependence on the perchlorate ion concentration for ClO_4^-/Cl^- solutions $\mu = 2{\cdot}0$ M, and reaction which occurs (with $[Eu^{II}] = 0{\cdot}04$ M, $[H^+] = 0{\cdot}5$ N and $[ClO_4^-] = 1{\cdot}95$ M, there is ~14% reaction of the Eu^{II} over 10 days) is believed to be due to the reaction of Eu^{II} with traces of O_2. With V^{II}, V^{III} and Ti^{III} reactions are sufficiently fast at temperatures of around 45°C for full kinetic studies to be possible. In each case the overall reaction is of the type

$$8Ti^{III} + ClO_4^- + 8H^+ \rightarrow 8Ti^{IV} + Cl^- + 4H_2O,$$

there being a one-equivalent oxidation of the metal ions. There is a simple first-order dependence on the two principal reactants, e.g.

$$-\frac{-d[Ti^{III}]}{dt} = k_1[Ti^{III}][ClO_4^-],$$

3. A. ADIN and A. G. SYKES, *Nature*, **209**, 804(1966), and *J. Chem. Soc.* (A), 1230 (1966).

and the first step is, therefore, rate determining, subsequent reactions being fast.

The reaction of V^{II} is complicated by the fact that the V^{III} reacts at a comparable rate with perchlorate,[3a] the V^{IV} which is formed reacting rapidly with V^{II},

$$V^{IV} + V^{II} \xrightarrow{\text{fast}} 2V^{III}.$$

The overall rate equation is of the form

$$-d[V^{II}]/dt = k_2[V^{II}][ClO_4^-] + 2k_3[V^{III}][ClO_4^-],$$

where k_3 can be obtained from a separate study of the reaction of V^{III} with perchlorate,

$$-d[V^{III}]/dt = k_3[V^{III}][ClO_4^-].$$

In the reaction with Ti^{III} there is a direct hydrogen-ion dependence,[4] but the addition of chloride retards the reaction apparently due to the formation of $TiCl^{2+}$. The vanadium reactions also show a small hydrogen-ion dependence, but the exact nature of this is not clear and medium effects may be relevant. In these reactions k_1, k_2 and k_3 are of the order 10^{-3}, 10^{-5} and 10^{-6} l mole^{-1} sec^{-1} respectively.

The product resulting from the reduction of the perchlorate in the rate determining first step is uncertain. A mechanism involving electron-transfer from the metal ion to a *d*-orbital on the chlorine atom of the perchlorate ion is perhaps the most reasonable. A two-equivalent change resulting from the transfer of an oxygen-atom to the metal ion, although clearly a possibility with V^{II} and V^{III} (since two-equivalent changes are possible) is hardly feasible with Ti^{III}. The only other alternative involves the transfer of an O^- ion from the perchlorate to the metal ion.

3a. W. B. King and C. S. Garner, *J. Phys. Chem.* **58**, 29 (1954).
4. F. R. Duke and P. R. Quinney, *J. Am. Chem. Soc.* **76**, 3800 (1954).

The reaction of chromium(II) with halogens

The oxidation of Cr^{2+} solutions with halogens provides with bromine and iodine a convenient means of preparing the 1 : 1 chromic halide complexes. In a typical case reaction solutions at 20°C were 1 N in $HClO_4$,[5] and except with iodine there were no halide ions present initially. The halogens were added in excess and after completion of the reaction the unreacted chlorine and bromine were removed by passing a stream of air through the solution. Excess iodine can be reduced to iodide by passing sulphur dioxide through the reacted solution, thus permitting spectrophotometric measurements.

In the reaction with chlorine only 75% of the chloride produced is complexed to the Cr^{III} product, while, with bromine, there is no free bromide immediately after the reaction. The direct test for free iodide is not valid (since iodide ions are present initially), but other evidence suggests that this reaction also proceeds with the retention of the iodide ion in the inner coordination sphere of the Cr^{III}. The reaction with chlorine is believed to proceed, in part, by an initial two-electron transfer in which Cr^{IV} is first produced,

$$Cr^{2+} + Cl_2 \rightarrow CrCl^{3+} + Cl^-.$$

The Cr^{IV} is probably substitution labile,

$$CrCl^{3+} \rightleftharpoons Cr^{IV} + Cl^-,$$

so that subsequent reaction with Cr^{II} to give Cr^{III} may or may not be with retention of the chloride. With the less powerful oxidizing agents, bromine and iodine, the reaction can proceed by the one-electron path only. This may be written

$$Cr^{2+} + X_2 \rightarrow CrX^{2+} + X$$

$$Cr^{2+} + X \xrightarrow{\text{fast}} CrX^{2+},$$

there being complete retention of both halide ions.

---

5. H. Taube and H. Myers, *J. Am. Chem. Soc.* 76, 2106 (1954).

Oxidation of metal ions with molecular oxygen

The oxidation of Fe^{II} with molecular oxygen in perchloric acid solutions,

$$4Fe^{2+} + O_2 + 4H^+ \rightarrow 4Fe^{3+} + 2H_2O,$$

has been studied over a temperature range 25–40°C.[6] In such experiments the gas is generally passed through stirred reaction solutions for at least ten minutes before the metal ion is added, and then subsequently, for the duration of the experiment. At various pHs and ionic strengths (adjusted with sodium perchlorate) and with oxygen first of all at atmospheric pressure, and then at 100–130 atm (when a special apparatus is used), the same kinetics are obtained,

$$\frac{-d[Fe^{2+}]}{dt} = k_{obs}[Fe^{2+}]^2[O_2].$$

Hydrogen peroxide, or peroxide derivatives, are believed to be formed,

$$Fe^{2+} + O_2 \overset{\text{fast}}{\rightleftharpoons} Fe^2 + O_2 \qquad (1)$$

$$Fe^{2+}O_2 + Fe^{2+} + 2H^+ \rightarrow 2Fe^{3+} + H_2O_2, \qquad (2)$$

two more ferrous ions reacting with the peroxide, and thus completing the reduction of oxygen to water,

$$2Fe^{2+} + H_2O_2 + 2H^+ \overset{\text{fast}}{\rightarrow} 2Fe^{3+} + 2H_2O. \qquad (3)$$

In this reaction sequence, (2) has to be modified, however, since k_{obs} is not directly proportional to $[H^+]^2$ as suggested here, but is almost independent of the hydrogen-ion concentration. Alternatively, it can be written

$$Fe^{2+}O_2 + H_2OFe^{2+} \rightarrow FeO_2H^{2+} + HOFe^{2+},$$

the peroxide species FeO_2H^{2+} reacting with further quantities of Fe^{2+} as in (3). The small inverse acid dependence which is

---

6. P. George, *J. Chem. Soc.* 4349 (1954).

actually observed is attributed to the ionization of a water molecule which solvates the ferrous–oxygen complex. Other evidence for the existence of FeO_2H^{2+} has been obtained in the reaction of Fe^{2+} with hydrogen peroxide.

The oxidation of Pu^{III} to Pu^{IV} has been studied with oxygen at atmospheric pressure. The reaction is very slow in solutions of hydrochloric acid and perchloric acid, but in aqueous sulphate solutions, $\mu = 2{\cdot}0$, it is measurably fast,[7]

$$4Pu^{III} + O_2 + 4H^+ \rightarrow 4Pu^{IV} + 2H_2O.$$

At 25°C the initial rate is second order in Pu^{III} and first order in oxygen. Since the reaction is essentially zero order in hydrogen ions, the mechanism is probably similar to that in the reaction with Fe^{II}. The dependence on sulphate suggests that there are two paths, one involving two and the other three sulphate ions in the activated complex.

In the oxidation of V^{III} to V^{IV} the rate equation is first order in both reactants,[8]

$$d[V^{III}]/dt = k_{obs}[V^{III}][O_2],$$

and V^V and HO_2^- are probably formed as intermediates, i.e. there is a two equivalent change. The V^V reacts rapidly with V^{III}

$$V^{III} + V^V \xrightarrow{\text{fast}} 2V^{IV}$$

and HO_2^- with two further V^{III} ions. With Cu^{II} present as catalyst, the reaction is independent of oxygen and the rate determining step is

$$V^{III} + Cu^{II} \rightarrow V^{IV} + Cu^I.$$

Subsequent reactions of Cu^I with oxygen are rapid, a cuprous–oxygen complex being formed in the first instance,

$$Cu^+ + O_2 \rightarrow Cu^+O_2.$$

7. T. W. Newton and F. B. Baker, *J. Phys. Chem.* **60**, 1417 (1956).
8. J. B. Ramsey, R. Sugimoto and H. DeVorkin, *J. Am. Chem. Soc.* **63**, 3480 (1941).

Although U^{IV} is a two equivalent reductant, the oxidation of U^{IV} to U^{VI},

$$-d[U^{IV}]/dt = k_{obs}[U^{IV}][O_2],$$

is not accomplished in a single two equivalent step.[9] It has been shown that reaction occurs by a chain mechanism:

Initiation: $U^{IV} + O_2 \rightarrow U^{V} + HO_2,$

Propagation: $U^{V} + O_2 \rightarrow U^{VI} + HO_2,$

$U^{IV} + HO_2 \rightarrow U^{V} + H_2O_2,$

Termination: $U^{V} + HO_2 \rightarrow U^{VI} + H_2O_2.$

This mechanism involves the formation of HO_2 and U^{V} as unstable intermediates, but is nevertheless favoured over the direct two-equivalent change, because of the " amplification factor " introduced by the chain reaction.

When chromous perchlorate solutions come into contact with atmospheric oxygen, there is a fast reaction ($k > 7 \times 10^5$ l mole^{-1} sec^{-1}) and a green dinuclear Cr^{III} compound (as distinct from the blue mononuclear Cr^{III}) is formed. Ion exchange experiments have indicated that this has a +4 charge in which case it could be either

$$(H_2O)_5Cr\text{—}O\text{—}Cr(H_2O)_5^{4+}$$

or

$$(H_2O)_4Cr\langle{}^{OH}_{OH}\rangle Cr(H_2O)_4^{4+}.$$

Oxygen-18 exchange studies have shown that there are 5 and not 5·5 oxygen atoms per chromium atom, thus indicating the second of these structures.[10] Isotopic labelling of the oxygen gas has furthermore shown that both the oxygen atoms are transferred to the product ion. These results can be understood in terms of

9. J. Halpern and J. G. Smith, *Canad. J. Chem.* **34**, 1419 (1956).
10. R. W. Kalaczkowski and R. A. Plane, *Inorg. Chem.* **3**, 322 (1964).

a mechanism which involves the formation of a peroxo-bridged intermediate. Thus the overall reaction may be written

$$4Cr(H_2O)_6^{2+} + O_2 \rightarrow 2(H_2O)_4Cr\langle{}^{OH}_{OH}\rangle Cr(H_2O)_4^{4+} + 6H_2O,$$

and the kinetics have been reported as showing a second-order dependence on the Cr^{2+} and a first-order dependence on oxygen.

Reduction of metal ions by molecular hydrogen

Molecular hydrogen reacts homogeneously with a number of metal ions in aqueous solution, reducing them to the metal or to a lower oxidation state. The kinetics of the primary (rate deter-

TABLE 27

REDUCTION OF METAL IONS BY HYDROGEN

Ion	Rate	$\Delta H^‡$ (kcal)	$\Delta S^‡$ (e.u.)
Cu^{2+}	$k[H_2][Cu^{2+}]$	26·6	−10
Hg^{2+}	$k[H_2][Hg^{2+}]$	18·1	−12
$Hg_2{}^{2+}$	$k[H_2][Hg_2{}^{2+}]$	20·4	−10
Ag^+	$k[H_2][Ag^+]$	24·0	−6
	$k[H_2][Ag^+]^2$	14·7	−25
$MnO_4{}^-$	$k[H_2][MnO_4{}^-]$	14·7	−17

mining) steps in these reactions are shown in Table 27. Since the formation of hydrogen atoms or H_2^+ ions is energetically unfavourable, hydrogen might be expected to behave as a two-equivalent reductant. The reactions with Hg^{2+}, Hg_2^{2+} and MnO_4^- are consistent with this. Thus the mechanism for the reduction of Hg^{2+} to Hg_2^{2+} is probably[11]

$$Hg^{2+} + H_2 \rightarrow Hg^0 + 2H^+$$

$$Hg^0 + Hg^{2+} \xrightarrow{\text{fast}} Hg_2^{2+}.$$

11. G. J. KORINEK and J. HALPERN, *J. Phys. Chem.* **60**, 285 (1956).

In the reactions with Hg_2^{2+} and MnO_4^-, the rate determining step again appears to be a two-equivalent reduction, leading to the formation of Hg_2^0 and Mn^V, respectively, as intermediates.

The reactions with Cu^{2+} and Ag^+ follow different paths, since mechanisms involving rate determining steps of the type

$$Cu^{2+} + H_2 \rightarrow Cu^+ + H^+ + H \quad (\Delta H \sim 54 \text{ kcal})$$

or

$$Cu^{2+} + H_2 \rightarrow Cu + 2H^+, \qquad (\Delta H \sim 66 \text{ kcal})$$

are energetically inconsistent with the observed activation energy of, in this case, 26 kcal. The mechanism of the reduction of Cu^{2+} to Cu^+ has been shown to be[12]

$$Cu^{2+} + H_2 \rightleftharpoons CuH^+ + H^+,$$

$$CuH^+ + Cu^{2+} \xrightarrow{\text{fast}} 2Cu^+ + H^+,$$

the intermediate CuH^+ being stabilized by covalent bonding. In the reduction of Ag^+ by H_2 there are two rate terms corresponding to reactions[13]

$$Ag^+ + H_2 \rightarrow AgH + H^+,$$

$$2Ag^+ + H_2 \rightarrow 2AgH^+.$$

The first of these probably involves formation of a covalently bonded hydride intermediate as in the reaction of Cu^{2+}. The second, on the other hand, obeys third-order kinetics and could well be a termolecular process. A similar termolecular rate determining step is observed in the Ag^+ catalysed reduction of MnO_4^- by H_2,

$$Ag^+ + MnO_4^- + H_2 \rightarrow AgH + MnO_4^{2-} + H^+.$$

Silver(I) also catalyses the reduction of both $Cr_2O_7^{2-}$ and Ce^{4+} with H_2.

---

12. E. Peters and J. Halpern, *J. Phys. Chem.* **59**, 793 (1955).
13. A. H. Webster and J. Halpern, *J. Phys. Chem.* **61**, 1239, 1245 (1957)

CHAPTER 11

REACTIONS OF THE HYDRATED ELECTRON

THE existence of the solvated electron in solutions of alkali metals in liquid ammonia, methylamine and ethylamine, has been recognized for some time. More recently (1962), direct evidence was obtained for the hydrated electron,[1] and solutions of up to 10^{-3} M concentrations can be prepared by subjecting de-aerated water to a high energy source, e.g. electrons from a Van de Graaf generator. In a typical procedure, the 6×1 cm face of a 6 cm^3 quartz cell (containing the sample of water) is uniformly irradiated with a 2×10^{-4} sec pulse of electrons from a 1·8 MeV linear accelerator. The solutions obtained have only a brief transitory existence, however, since the hydrated electron reacts rapidly with H^+ ions which are present in solution. The spectrum (Fig. 28) can be recorded using sensitive photomultiplier/oscilloscope techniques and is similar to that observed for solutions of alkali metals in liquid ammonia.

The orbital containing the electron is thought to extend over not just one but several water molecules. It can therefore be represented by a formula $(H_2O)_n^-$ or, since n is not known, more simply e_{aq}^-. In its reactions, the hydrated electron obeys the Brønsted–Bjerrum Theory for ionic reactions.[2] Thus, in the reaction with, for example, Ag^+, a plot of log k against $\mu^{1/2}$,

1. E. J. HART and J. W. BOAG, *J. Am. Chem. Soc.* **84**, 4090 (1962).
2. E. COLLINSON, F. S. DAINTON, D. R. SMITH and S. TAZUKE, *Proc. Chem. Soc.* 140 (1962).

where μ is the ionic strength, gives a straight line of gradient $-z_1z_2 = 1$. The rate constant for the reaction with water,

$$e^-_{aq} + H_2O \xrightarrow{k} H + OH^-,$$

is relatively slow, $k = 2 \cdot 7 \times 10^4\ sec^{-1}$.

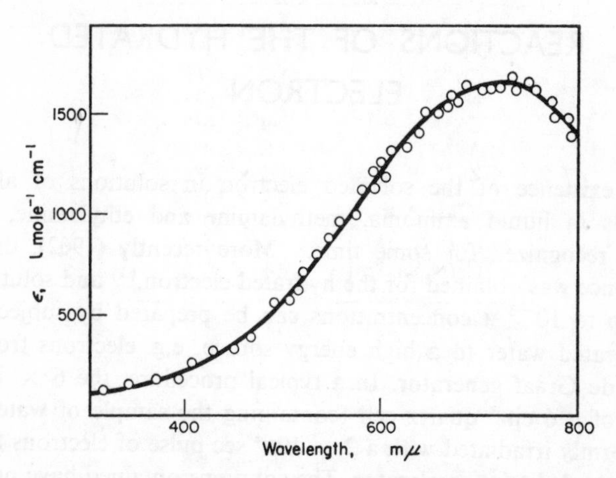

FIG. 28. Absorption spectrum of the hydrated electron in pure water. [Reproduced, with permission, from J. H. BAXENDALE *et al.*, *Nature*, **201**, 468 (1964).]

Reactions with metal ions

Reactions of the hydrated electron can be studied by having solutes present in the irradiated sample of water. In the presence of solute, the decay of the 700 mμ hydrated electron peak is much more rapid than would otherwise be observed. With concentrations of metal ions in the 0·005–0·05 M region and small hydrogen-ion concentrations, reactions which the hydrated electron would normally undergo in pure water,

$$e^-_{aq} + H_2O \rightarrow H + OH^-$$

and

$$e^-_{aq} + H^+ \rightarrow H,$$

TABLE 28

RATE CONSTANTS FOR THE REACTIONS OF e_{aq}^{-} WITH SOME METAL IONS AT ROOM TEMPERATURE AND pH 7

Reactant	k (l mole^{-1} sec^{-1})
Mn^{2+}	$4{\cdot}0 \times 10^{7}$
Fe^{2+}	$1{\cdot}6 \times 10^{8}$
Co^{2+}	$1{\cdot}2 \times 10^{10}$
Ni^{2+}	$2{\cdot}2 \times 10^{10}$
Cu^{2+}	$2{\cdot}9 \times 10^{10}$
Ag^{+}	$3{\cdot}2 \times 10^{10}$
Zn^{2+}	$1{\cdot}5 \times 10^{9}$
Cd^{2+}	$5{\cdot}2 \times 10^{10}$
Pb^{2+}	$3{\cdot}9 \times 10^{10}$
$Co(NH_3)_6^{3+}$	9×10^{10}
$Co(NO_2)_6^{3-}$	$5{\cdot}8 \times 10^{10}$

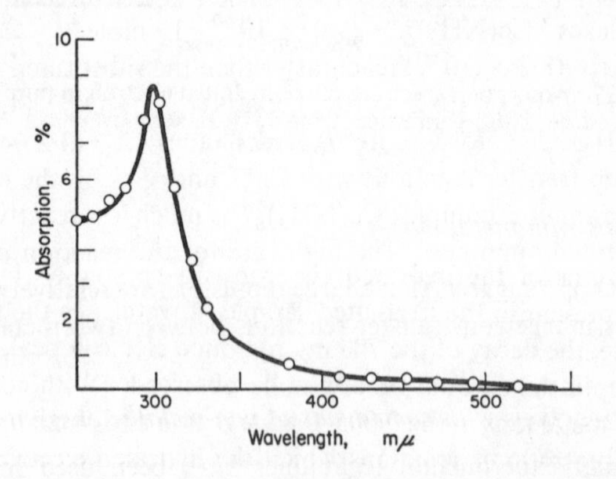

FIG. 29. The transient absorption spectrum in $5{\cdot}3 \times 10^{-5}$ M $NiSO_4$. The spectrum was taken 2×10^{-4} sec after an electron pulse and is attributed to Ni^{+}. [Reproduced, with permission, from J. H. BAXENDALE, E. M. FIELDEN and J. P. KEENE, *Proc. Chem. Soc.* 243 (1963).]

are negligible, and rate constants can be obtained from the observed rate of decay[3] (Table 28).

Transient absorption spectra which have been observed with Mn^{2+}, Co^{2+}, Ni^{2+}, Zn^{2+}, Cd^{2+} and Pb^{2+}, in the 300 mμ region, have been attributed to the formation of unstable oxidation states. Thus in the reaction with Ni^{2+},

$$Ni^{2+} + e_{aq}^{-} \rightarrow Ni^{+},$$

the transient spectrum shown in Fig. 29 is believed to be due to the Ni^{+} ion. Metal ions which readily hydrolyse are more difficult to study since higher hydrogen-ion concentrations are necessary, and, under these conditions, the reaction of the hydrated electron with H^{+} is much more effective. No clear picture has yet emerged from these experiments as to relative ease with which the t_{2g} and e_g orbitals on a metal ion accept electrons. It has been noted, however, that the activation energy for the reaction with Mn^{2+} (4·0 kcal mole^{-1}) is higher than with other metal ions, and about twice that for the reaction with Co^{2+}.

Reactions with cobalt(III) complexes are of interest in that the complexes $Co(NH_3)_6^{3+}$ ($9{\cdot}0 \times 10^{10}$ 1 mole^{-1} sec^{-1}) and $Co(en)_3^{3+}$ ($8{\cdot}2 \times 10^{10}$) react faster than the substituted complexes $Co(NH_3)_5H_2O^{3+}$ ($6{\cdot}2 \times 10^{10}$), $Co(NH_3)_4(H_2O)_2^{3+}$ (4.4×10^{10}) $Co(NH_3)_5Cl^{2+}$ ($5{\cdot}4 \times 10^{10}$) and $Co(en)_2Cl_2^{+}$ ($3{\cdot}2 \times 10^{10}$). In electron-transfer reactions with Cr^{2+} and V^{2+} on the other hand the hexammine complex $Co(NH_3)_6^{3+}$ is much less reactive than the substituted ammines. The high rate for the reaction of e_{aq}^{-} with $Co(NO_2)_6^{3-}$ suggests that charge repulsions are relatively unimportant as in electron-transfer reactions between two metal ions.

Some reactions with compounds of non-metallic elements

Similar experimental techniques have been used in studying

---

3. J. H. Baxendale, E. M. Fielden, C. Capellos, J. M. Francis, J. V. Davies, M. Ebert, C. W. Gilbert, J. P. Keene, E. J. Land, A. J. Swallow and J. M. Nosworthy, *Nature* **201**, 468 (1964), and J. H. Baxendale, E. M. Fielden and J. P. Keene, *Proc. Roy. Soc.* 286A, 322 (1965).

reactions of the hydrated electron with non-metallic solutes[4] (Table 29). Known electron scavengers such as the proton, hydrogen peroxide, oxygen, nitric oxide, nitrous oxide and carbon dioxide have high rate constants, and organic scavengers such as acetone, chloroform and carbon tetrachloride also react rapidly. The capture of an electron may or may not involve dissociation of the reactant. Thus with, for example, oxygen, HO_2 radicals are formed,

$$e_{aq}^- + O_2 \rightarrow O_2^- \xrightarrow{H^+} HO_2,$$

and, similarly, with carbon dioxide, COOH radicals are obtained without dissociation of the reactant,

$$e_{aq}^- + CO_2 \rightarrow CO_2^- \xrightarrow{H^+} COOH.$$

TABLE 29

RATE CONSTANTS FOR REACTIONS OF e_{aq}^- WITH SOME NON-METALLIC SOLUTES AT ROOM TEMPERATURE

Reactant	k (l mole^{-1} sec^{-1})
H^+	$2{\cdot}4 \times 10^{10}$
H_2O_2	$1{\cdot}2 \times 10^{10}$
O_2	$1{\cdot}8 \times 10^{10}$
NO	$3{\cdot}1 \times 10^{10}$
N_2O	$8{\cdot}7 \times 10^{9}$
CO_2	$7{\cdot}7 \times 10^{9}$
NO_3^-	$1{\cdot}1 \times 10^{10}$
SO_4^{2-}	$<10^6$
ClO_4^-	$<10^6$
HAc	$\sim 10^8$
Ac^-	$<10^6$

With nitrous oxide and hydrogen peroxide on the other hand there is dissociative capture of the hydrated electron,

$$e_{aq}^- + N_2O \rightarrow N_2 + O^- \xrightarrow{H^+} OH$$

4. S. GORDON, E. J. HART, M. S. MATHESON, J. RABANI and J. K. THOMAS, *Discussions Faraday Soc.* **36**, 793 (1963).

and

$$e^-_{aq} + H_2O_2 \rightarrow OH + OH^-.$$

In all these systems, the initial steps, as shown here, are rate determining.

It is of particular interest to note that the reaction

$$H^+ + OH^- \rightarrow H_2O, \ (k = 1{\cdot}4 \times 10^{11}\,\text{l mole}^{-1}\,\text{sec}^{-1})$$

involving the transfer of a proton from H_3O^+ to OH^-, is some six times faster than the reaction of e^-_{aq} with H^+, i.e. H_3O^+. The activation energy of the latter has been measured at a hydrogen-ion concentration of $4{\cdot}4 \times 10^{-5}$ N and found to be 3·2 kcal mole^{-1}.[5] This may be compared with the value 2–3 kcal mole^{-1} found for the reaction of H^+ and OH^-. Since rate constants for these reactions must be close to the diffusion controlled limit, these results suggest that 2–3 kcal mole^{-1} is the activation energy required for the diffusion of the above species in water. For the reaction

$$e^-_{aq} + H_2O \rightarrow H + OH^-, \ (k = 2{\cdot}3 \times 10^4\,\text{l mole}^{-1}\,\text{sec}^{-1})$$

the activation energy is 4·6 kcal mole^{-1}.

That the solvated electron should be more stable in liquid ammonia than in water deserves some further comment. The reaction corresponding to that of e^-_{aq} with H_3O^+, namely, $e^-_{NH_3}$ with NH_4^+, is not as effective, since the concentration of NH_4^+ in liquid ammonia is much smaller. Thus, the equilibrium constant for the ionization

$$2NH_3 \rightleftharpoons NH_4^+ + NH_2^-,$$

is $\sim 10^{-30}\,\text{l}^{-2}\,\text{mole}^2$ at -50°C, and in neutral solutions of liquid ammonia the concentration of NH_4^+ is approximately $10^{-15}\,\text{l}^{-1}$ mole.

5. J. K. Thomas, S. Gordon and E. J. Hart, *J. Phys. Chem.* **68,** 1524 (1964).

Electron-transfer reactions; The hydrated electron is not *formed as an intermediate*

An alternative mechanism for electron-transfer reactions between metal ions is one in which the electron spends sufficient time in solution, in transit from reactant A to B, to become solvated. Thus, for the one-equivalent reaction of Cr^{II} with Fe^{III}, a possible reaction sequence is

$$Cr^{II} \underset{k_{-1}}{\overset{k_1}{\rightleftharpoons}} Cr^{III} + e^-_{aq}$$

$$Fe^{III} + e^-_{aq} \xrightarrow{k_2} Fe^{II},$$

and making the stationary-state assumption $d[e^-_{aq}]/dt = 0$, a rate equation

$$\frac{-d[Fe^{III}]}{dt} = \frac{k_1k_2[Fe^{III}][Cr^{II}]}{k_{-1}[Cr^{III}] + k_2[Fe^{III}]},$$

is obtained. Two limiting cases of this expression are of interest, firstly, with $k_2[Fe^{III}] \gg k_{-1}[Cr^{III}]$ when the above equation approximates to

$$-d[Fe^{III}]/dt = k_1[Cr^{II}],$$

and, secondly, with $k_{-1}[Cr^{III}] \gg k_2[Fe^{III}]$, when

$$\frac{-d[Fe^{III}]}{dt} = \frac{k_1k_2[Fe^{III}][Cr^{II}]}{k_{-1}[Cr^{III}]}.$$

Since neither of these limiting cases, nor the original rate equation, are obeyed by this or any other system, clearly there is no evidence for intermediate formation of the hydrated electron in electron-transfer reactions.

CHAPTER 12

SUBSTITUTION REACTIONS OF OCTAHEDRAL COMPLEXES

In the first transition series, ligand substitution reactions of octahedral complexes of $Cr^{III}(3d^3)$ and spin-paired $Co^{III}(3d^6)$ (and to a lesser extent $Ni^{II}(3d^8)$) are sufficiently slow to be measured by conventional means. Such complexes are said to be inert or non-labile. The slowness of substitution is attributable to the high crystal-field stabilization energy which the t_{2g}^3 and t_{2g}^6 3+ ions (and to a lesser extent $(t_{2g}^6 e_g^2)$) forms have. As a result of this stabilization, the activation energies required to form reaction intermediates are high and rates slow (see, for example, Table 30 below). In the first transition series, Fe^{II} also has a d^6 electron configuration, but it generally forms high-spin complexes which have little crystal-field stabilization and which are therefore labile. Exceptions are the $Fe(CN)_6^{4-}$, $Fe(bipy)_3^{2+}$ and $Fe(phen)_3^{2+}$ complexes, in which the crystal-field provided by the ligands is sufficiently strong to cause electron pairing. Complexes of second and third transition series elements are generally low-spin and substitution inert.

The relative ease with which a large number of Co^{III} and, to a lesser extent, Cr^{III} complexes can be prepared has, until recently, tended to confine kinetic studies to these forms. With the advent of fast reaction techniques, it is now possible to measure the rate of substitution of the inner coordination sphere of labile metal ions (see, for example, the kinetic data in Table 2, p. 37).

Acid hydrolysis of cobalt(III) pentammine complexes

A group of substitution reactions which have been extensively studied are those involving hydrolysis of pentamminecobalt(III) complexes $Co^{III}(NH_3)_5X^{2+}$. In acid solutions, pH < 3, the substitution may be written

$$Co^{III}(NH_3)_5X^{2+} + H_2O \rightarrow Co^{III}(NH_3)_5H_2O^{3+} + X^-,$$

TABLE 30

KINETIC DATA FOR THE ACID HYDROLYSIS OF SOME PENTAMMINE-COBALT(III) COMPLEXES $Co(NH_3)_5X^{2+}$ AT 25°C

Complex	k_a (sec^{-1})	$\Delta H^\ddagger$ (kcal mole^{-1})
$Co(NH_3)_5NO_3{}^{2+}$	$2{\cdot}6 \times 10^{-5}$	26
$Co(NH_3)_5I^{2+}$	$8{\cdot}3 \times 10^{-5}$	—
$Co(NH_3)_5H_2O^{3+}$	$5{\cdot}8 \times 10^{-6}$	27
$Co(NH_3)_5Br^{2+}$	$6{\cdot}3 \times 10^{-6}$	24
$Co(NH_3)_5Cl^{2+}$	$1{\cdot}6 \times 10^{-6}$	23
$Co(NH_3)_5SO_4{}^{+}$	$1{\cdot}16 \times 10^{-6}$	19
$Co(NH_3)_5Ac^{2+}$	$1{\cdot}2 \times 10^{-7}$	—
$Co(NH_3)_5NCS^{2+}$	$5{\cdot}0 \times 10^{-10}$	31
$Co(NH_3)_5NO_2{}^{2+}$	very slow	—
$Co(NH_3)_5N_3{}^{2+}$	$2{\cdot}1 \times 10^{-9}$	34

where, in the majority of cases, X^- is a singly charged anion There is no $[H^+]$ dependence however, and since water is in large excess, reactions show a first-order dependence on the complex only,(1) i.e.

$$-d[Co(NH_3)_5X^{2+}]/dt = k_a[Co(NH_3)_5X^{2+}].$$

Rate constants obtained for these reactions (Table 30), show a linear dependence on thermodynamic stability (plot of log k_a against log K). Thus the nitrate complex (or Co–NO_3 bond) is the least stable of those listed in Table 30.

1. See, e.g. A. W. ADAMSON and R. G. WILKINS, *J. Am. Chem. Soc.* **76**, 3379 (1954).

The two limiting mechanisms for substitution of octahedral complexes are generally described as S_N1 and S_N2 processes, respectively (substitution, nucleophilic and either first or second order). They are, for the acid hydrolysis of $Co(NH_3)_5X^{2+}$ complexes,

$$\left.\begin{array}{l} Co(NH_3)_5X^{2+} \rightarrow Co(NH_3)_5^{3+} + X^- \\ Co(NH_3)_5^{3+} + H_2O \xrightarrow{\text{fast}} Co(NH_3)_5H_2O^{3+} \end{array}\right\} S_N1 \quad (1)$$

$$\left.\begin{array}{l} Co(NH_3)_5X^{2+} + H_2O \rightarrow Co(NH_3)_5XH_2O^{2+} \\ Co(NH_3)_5XH_2O^{2+} \xrightarrow{\text{fast}} Co(NH_3)_5H_2O^{3+} + X^-. \end{array}\right\} S_N2 \quad (2)$$

In (1), an unstable five-coordinate intermediate, and in (2), an unstable seven-coordinate intermediate is formed. Clearly, the route a reaction follows need not necessarily conform to one or other of these extreme cases, but, by reference to these, the order in which bond-making and bond-breaking takes place can be indicated. Thus, to say a reaction has S_N1 character, would imply that bond stretching is of first importance, although a water molecule may move into the Co^{III} inner-coordination sphere before a five-coordinate intermediate is actually formed. Since water is in large excess, both (1) and (2) are first order in complex, and a direct kinetic distinction between the two mechanisms is not possible.

For many reactions there would appear to be a balance between S_N1 and S_N2 character depending on the nature of the incoming and outgoing groups, their size and nucleophilic character, for example. Information regarding the manner of the substitution,

$$Co(NH_3)_5H_2O^{3+} + H_2{}^{18}O \rightarrow Co(NH_3)_5H_2{}^{18}O^{3+} + H_2O,$$

has been obtained by studying the effect which pressures of up to 7000 atm have on the reaction.[2] With increasing pressure, a small but definite decrease in rate was observed. The difference

2. H. R. Hunt and H. Taube, *J. Am. Chem. Soc.* **80**, 2642 (1958).

in volume between the activated complex and reactants, $\Delta V^{\ddagger}$, can be calculated using the expression

$$\Delta V^{\ddagger} = \frac{-RT\ln(k_2/k_1)}{p_2 - p_1},$$

where k_1 and k_2 are the rate constants at pressures p_1 and p_2, respectively. Over the pressure range studied, $\Delta V^{\ddagger}$ was found to be $1{\cdot}2 \pm 0{\cdot}2$ ml. It can be shown that for an S_N1 process

$$Co(NH_3)_5H_2O^{3+} \rightarrow Co(NH_3)_5^{3+} + H_2O,$$

$\Delta V^{\ddagger}$ should be *ca.* 3·6 ml, and for an S_N2 process

$$Co(NH_3)_5H_2O^{3+} + H_2O \rightarrow Co(NH_3)_5(H_2O)_2^{3+},$$

$\Delta V^{\ddagger}$ should be negative, and in the limiting case $-3{\cdot}6$ ml. The value obtained excludes a limiting S_N2 process, and is more in keeping with a modified S_N1 process in which the Co^{III}—OH_2 bond first stretches to a definite critical distance before the incoming water molecule begins to move in.

A reaction in which a limiting S_N1 mechanism seems to be applicable is the Hg^{II} catalysed hydrolysis of the halo complexes $Co(NH_3)_5X^{2+}$.[2a] The method used to test for $Co(NH_3)_5^{3+}$ was to utilize the expected isotopic discrimination in the eventual reaction with water (which is enriched in oxygen-18),

$$Co(NH_3)_5^{3+} + H_2O \rightarrow Co(NH_3)_5H_2O,$$

with Hg^{II} the ratio of $H_2^{16}O$ and $H_2^{18}O$ in the product is the same for each halo complex. With Ag^{I} and Tl^{III} on the other hand the ratios vary and a limiting S_N1 mechanism may not be applicable.

The slower rate constants for the acid hydrolysis of $Rh(NH_3)_5Br^{2+}$ ($k \sim 2 \times 10^{-7}$ sec^{-1}) and $Ir(NH_3)_5Br^{2+}$ ($k \sim 2 \times 10^{-9}$ sec^{-1}), are in keeping with the increased crystal-field stabilization for these d^6 ions. With $Cr^{III}(NH_3)_5X^{2+}$ complexes rates are faster and activation energies some 2 kcal mole^{-1} smaller than those for the corresponding Co^{III} complexes. Again this is as expected from a consideration of crystal-field effects.

2a. F. A. Posey and H. Taube, *J. Am. Chem. Soc.* **79**, 255 (1957).

Reactions involving only partial replacement of ligand groups in substitution complexes can occur more readily than those involving a complete replacement. Thus, when an aquo complex is placed in D_2O the exchange of hydrogen for deuterium atoms (without the exchange of oxygen atoms) is rapid

$$Co(NH_3)_4(H_2O)_2^{3+} \xrightarrow{D_2O} Co(NH_3)_4(D_2O)_2^{3+},$$

and intermediate formation of $Co(NH_3)_4H_2O\,OH^{2+}$ seems likely. Similarly the hexammine cobalt(III) complex exchanges its hydrogen atoms at a measurable rate

$$Co(NH_3)_6^{3+} \xrightarrow{D_2O} Co(ND_3)_6^{3+}.$$

Base hydrolysis of cobalt(III) complexes

The first term in k_a in the rate equation

$$d[Co(NH_3)_5X^{2+}]/dt = k_a[Co(NH_3)_5X^{2+}] + k_b[Co(NH_3)_5X^{2+}][OH^-]$$

is dominant below pH 3 as has already been indicated, while the second term corresponding to base hydrolysis becomes effective at higher pH values. For these systems which have been studied over the full pH range k_b values are found to be some 10^5 times greater than k_a, and with decreasing acid concentrations above pH 3, k_b quickly establishes itself as the dominant path. From the rate equation k_b might be a direct substitution by OH^-,

$$Co(NH_3)_5X^{2+} + OH^- \rightarrow Co(NH_3)_5OH^{2+} + X^-,$$

but since anion substitution reactions of the general type

$$Co(NH_3)_5X^{2+} + Y^- \rightarrow Co(NH_3)_5Y^{2+} + X^-,$$

are invariably slow, there is good reason for doubting this.

Experiments which have indicated a more likely mechanism for base hydrolysis are as follows. First of all, in the hydrolysis of $Co(NH_3)_4(H_2O)NO_3^{2+}$, the constant k_b is effective below pH 3.[3] This is thought to be due to participation by $Co(NH_3)_4(OH)NO_3^{2+}$, where small concentrations of the latter are formed in the rapid pH-controlled equilibrium

$$Co(NH_3)_4(H_2O)NO_3^{2+} \rightleftharpoons Co(NH_3)_4(OH)NO_3^{+} + H^+. \quad (1)$$

The rate determining steps are accordingly

$$Co(NH_3)_4(H_2O)NO_3^{2+} + H_2O \rightarrow Co(NH_3)_4(H_2O)_2^{3+} + NO_3^-, \quad (2)$$

which corresponds to a process k_a, and

$$Co(NH_3)_4(OH)NO_3^{+} + H_2O \rightarrow Co(NH_3)_4(OH)H_2O^{2+} + NO_3^- \quad (3)$$

which, together with the pre-equilibrium in (1), corresponds to base hydrolysis. An ammonia group in $Co(NH_3)_5X^{2+}$ is thought to behave in a similar though less pronounced way to the H_2O in $Co(NH_3)_4(H_2O)NO_3^{2+}$. The acidities of, for example, $Co(NH_3)_6^{3+}$, $Co(NH_3)_5Cl^{2+}$ and $Co(en)_3^{3+}$ are far too weak to detect in the normal way, but all three have been shown to exchange hydrogen atoms with D_2O, even in acid solution. The only reasonable way in which this might occur is with intermediate formation of an amido complex. That substitution of amido and hydroxo forms should be so favourable can be explained in terms of an S_N1 dissociation mechanism

$$Co(NH_3)_4(NH_2)X^+ \rightarrow Co(NH_3)_4(NH_2)^{2+} + X^-.$$

By π-bonding to the central metal atom,[4] the NH_2^- and OH^- groups are thought to both weaken the Co—X bond and stabilize the five-coordinate intermediate,

$$>\!\!Co\!<^{NH_2}_{X} \rightarrow\ >\!\!Co{=}NH_2 + X^-.$$

3. F. J. Garrick, *Nature* **139**, 507 (1937).
4. R. G. Pearson and F. Basolo, *J. Am. Chem. Soc.* **78**, 4878 (1956).

The full reaction sequence, which may be written

$$Co(NH_3)_5X^{2+} \overset{K}{\rightleftharpoons} Co(NH_3)_4(NH_2)X^+ + H^+ \quad \text{(fast)} \quad (4)$$

$$Co(NH_3)_4(NH_2)X^+ \overset{k'_b}{\rightarrow} Co(NH_3)_4(NH_2)^{2+} + X^- \quad (5)$$

(rate determining)

$$Co(NH_3)_4(NH_2)^{2+} + H_2O \rightarrow Co(NH_3)_5OH^{2+}, \quad \text{(fast)} \quad (6)$$

is known as the S_N1 conjugate-base (CB) mechanism. From (5), the rate may be expressed

$$-d[Co(NH_3)_5X^{2+}]/dt = k'_b[Co(NH_3)_4(NH_2)X^+],$$

and substituting for $[Co(NH_3)_4(NH_2)X^+]$ from (4), and for $[H^+]$ from the expression

$$K_w = [H^+][OH^-],$$

$$\frac{-d[Co(NH_3)_5X^{2+}]}{dt} = \frac{k'_bK}{K_w}[Co(NH_3)_5X^+][OH^-].$$

In this expression, K_w is $\sim 10^{-14}$ mole^2 l^{-2}, and since K is believed to be of the same order of magnitude, k'_b and the observed rate constant k_b are to a first approximation equal. In support of the S_N1 CB mechanism, it should be noted that hydrolysis of complexes $Co(CN)_5Br^{3-}$ and $Co(CN)_5I^{3-}$, which have no acid proton to lose, are independent of pH over the alkaline range.

In the hydrolysis of $Ru(NH_3)_5Cl^{2+}$ (a $4d^5$ ion), base hydrolysis has likewise been shown to be much faster than acid hydrolysis, the ratio of the rate constants being more than 10^6. With the corresponding chromium(III) ($3d^3$) and rhodium(III) ($4d^6$) complexes, on the other hand, the difference is considerably less, and base hydrolysis of $Cr(NH_3)_5X^{2+}$ complexes is not appreciable below pH 9. That base hydrolysis is much less effective for chromium(III) and rhodium(III) complexes is probably due to the decreased reactivity of conjugate-base forms.

The replacement of X^- by Y^- in $Co^{III}L_5X$ complexes

Few reactions are known in which the replacement of one negatively charged group by another takes place directly without the intermediate formation of a monoaquo complex. Substitution of X^- in $Co^{III}(CN)_5X^{3-}$ complexes appears to proceed by an S_N1 mechanism in aqueous solution,[5] and both H_2O and Y^- are thought to compete for the five-coordinate intermediate,

$$Co(CN)_5X^{3-} \rightleftharpoons Co(CN)_5^{2-} + X^-,$$

$$Co(CN)_5^{2-} + H_2O \rightarrow Co(CN)_5H_2O^{2-},$$

$$Co(CN)_5^{2-} + Y^- \rightarrow Co(CN)_5Y^{3-}.$$

In such systems, direct substitution of X^- by Y^- is possible.

In the majority of reactions, however, which are thought to have at least some S_N2 character, substitution of X^- by Y^- can take place only with the intermediate formation of an aquo complex, e.g.

$$Co(NH_3)_5X^{2+} + H_2O \rightleftharpoons Co(NH_3)_5H_2O^{3+} + X^-$$

$$Co(NH_3)_5H_2O^{3+} + Y^- \rightarrow Co(NH_3)_5Y^{2+} + H_2O.$$

In such reactions, the rate at which X^- is liberated is found to be zero order with respect to Y^-. The first example of a reaction of this type[6] was provided by the exchange of $^*Cl^-$ with *cis*–$Co(en)_2Cl_2^+$,

$$Co(en)_2Cl_2^+ + H_2O \rightarrow Co(en)_2(H_2O)Cl^{2+} + Cl^-$$

$$Co(en)_2(H_2O)Cl^{2+} + {}^*Cl^- \rightarrow Co(en)_2{}^*Cl_2^+ + H_2O,$$

when the amount of Cl^- liberated at any one time was found to be independent of the concentration of $^*Cl^-$. With $Co(NH_3)_5Cl^{2+}$, the rate of hydrolysis is similarly independent of sulphate, nitrate and perchlorate ion concentrations, except for small increases or decreases which can be attributed to ion-pair interactions $Co(NH_3)_5X^{2+}, Y^-$.

5. A. Haim and W. K. Wilmarth, *Inorg. Chem.* **1**, 583 (1962).
6. G. W. Ettle and C. H. Johnson, *J. Chem. Soc.* 1490 (1940).

The inability of Y^- to substitute directly an X^- ligand by a mechanism having some S_N2 character, is believed to be due to repulsion between the two groups in the transition complex.[7] Thus, for a *cis* attack, the transition complex is as shown in (I) below, and there is repulsion between X^- and Y^-. For *trans* attack in (II), on the other hand, much more reorganization is required and a greater loss of crystal-field stabilization energy is necessary (i.e. the activation energy is larger).

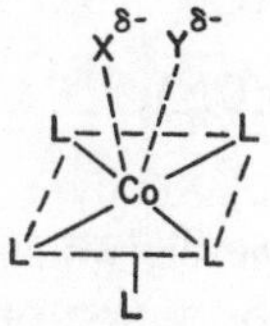

I 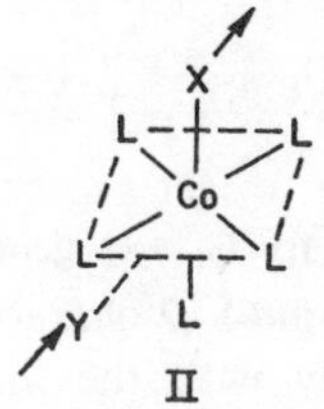 II

The replacement of H_2O by Y^- in $Co^{III}L_5H_2O$ complexes

Substitution reactions of the type

$$Co(NH_3)_5H_2O^{3+} + Y^- \rightarrow Co(NH_3)_5Y^{2+} + H_2O,$$

are found to be essentially first order in both the complex and substituting anion Y^- over a fairly wide range of conditions. Since the H_2O ligand can also be replaced by a solvent water molecule, Y^- and the solvent are in competition for a five-co-ordinate intermediate if an S_N1 mechanism is applicable. By using high concentrations of Y^-, it should, therefore, be possible to test for an S_N1 mechanism. Such experiments are not entirely satisfactory with positively charged complexes of the type $Co(NH_3)_5H_2O^{3+}$, because, in the first place, the rate of anation is slow compared to the rate of hydrolysis, and secondly, because ion-pair complexing is effective at high Y^- concentrations. To

7. See for example R. Dyke and W. C. E. Higginson, *J. Chem. Soc.* 2788 (1963).

avoid the latter, anionic complexes can be used, and in a recent study of some $Co(CN)_5H_2O^{2-}$ reactions,[8] it has been shown that these conform to a limiting S_N1 mechanism,

$$Co(CN)_5H_2O^{2-} \underset{k_{-1}}{\overset{k_1}{\rightleftharpoons}} Co(CN)_5^{2-} + H_2O$$

$$Co(CN)_5^{2-} + Y^- \xrightarrow{k_2} Co(CN)_5Y^{3-}.$$

Making the stationary-state approximation $d[Co(CN)_5^{2-}]/dt = 0$, a rate equation

$$\frac{d[Co(CN)_5H_2O^{2-}]}{dt} = \frac{k_1k_2[Co(CN)_5H_2O^{2-}][Y^-]}{k_{-1} + k_2[Y^-]}$$

is obtained. If Y^- is in large excess so that its concentration can be assumed constant for any one reaction, the reaction is pseudo-first order in complex

$$-d[Co(CN)_5H_2O^{2-}]/dt = k_{obs}[Co(CN)_5H_2O^{2-}],$$

where

$$k_{obs} = \frac{k_1k_2[Y^-]}{k_{-1} + k_2[Y^-]}.$$

The latter can be rearranged to give

$$\frac{1}{[Y^-]} = \frac{k_2k_1}{k_{-1}}\frac{1}{k_{obs}} - \frac{k_2}{k_{-1}},$$

and from plots of $1/[Y^-]$ against $1/k_{obs}$ for a series of runs with varying $[Y^-]$, k_2/k_{-1} and k_1 can be obtained. With different Y^- groups, k_1 values are in good agreement ($k_1 \sim 1{\cdot}6 \times 10^{-3}$ sec^{-1}) as is required by the above mechanism. Since k_{-1} is also independent of the substituent Y^-, the ratio of k_2/k_{-1} is a measure of the nucleophilic power of the different substituents. The order of k_2 values is accordingly

$$OH^- > N_3^- > NCS^- > I^- > Br^- > CNO^- > H_2O.$$

8. A. Haim and W. K. Wilmarth, *Inorg. Chem.* **1**, 573 (1962).

The acid hydrolysis of some cis *and* trans *cobalt(III) complexes*

A large variation in rate constant > 10^5 has been observed for a series of reactions,

$$Co^{III}(en)_2LCl^+ + H_2O \rightarrow Co^{III}(en)_2LH_2O^{2+} + Cl^-,$$

TABLE 31

RATE CONSTANTS (AT 25°C), STERIC COURSE, AND SUGGESTED MECHANISM FOR THE ACID HYDROLYSIS OF *cis*- AND *trans*-$Co^{III}(en)_2LCl^+$ COMPLEXES

cis L	k (sec^{-1})	% *cis* in product	Suggested mechanism
OH^-	$1{\cdot}2 \times 10^{-2}$	100%	S_N1
N_3^-	$2{\cdot}0 \times 10^{-4}$	100%	S_N1
Cl^-	$2{\cdot}4 \times 10^{-4}$	100%	S_N1
NCS^-	$1{\cdot}1 \times 10^{-5}$	100%	S_N1
NH_3	$\sim 5 \times 10^{-7}$	100%	S_N2
H_2O	$0{\cdot}95 \times 10^{-4}$	not known	S_N2
NO_2^-	$1{\cdot}1 \times 10^{-4}$	100%	S_N2
trans L			
OH^-	$1{\cdot}6 \times 10^{-3}$	75%	S_N1
N_3^-	$2{\cdot}2 \times 10^{-4}$	20%	S_N1
Cl^-	$3{\cdot}5 \times 10^{-5}$	25%	S_N'
NCS^-	$0{\cdot}5 \times 10^{-7}$	50–70%	S_N1
NH_3	4×10^{-8}	0	S_N2
NO_2^-	$0{\cdot}98 \times 10^{-3}$	0	S_N2

in which the ligand L and its position in the complex is varied[9] (Table 31). We note that with a number of these complexes, in particular with L = OH^-, Cl^- and NCS^-, the *cis* form reacts appreciably faster than the *trans* form, and L is said to labilize the *cis* positions in the complex (*cis* effect). For L = NO_2^- at least, the opposite is true, however, and the *trans* position is labilized. It is also of interest that, in the hydrolysis of *trans* complexes, a mixture of *cis* and *trans* isomers is obtained, except

9. M. E. BALDWIN, S. C. CHAN and M. L. TOBE, *J. Chem. Soc.* 4637 (1961), and references therein.

with L = NO_2^- and NH_3, when the original *trans* configuration is retained.

A reasonable explanation of these results is possible by considering the π-bonding properties of different L ligands, and mechanisms having S_N1 and S_N2 character have been assigned

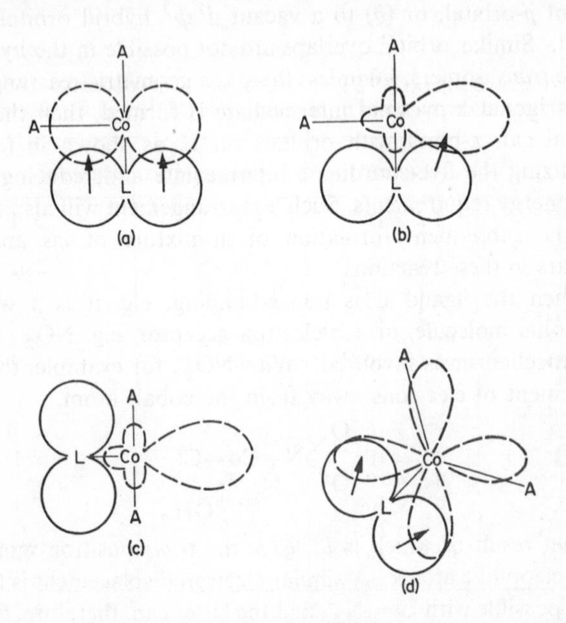

FIG. 30. The overlapping of a filled *p*-orbital of L with (*a*) a *p*-orbital, or (*b*) a d^2sp^3 hybrid orbital on the cobalt stabilizes the square pyramid intermediate resulting from the dissociation of chloride from the *cis*-$Co(A)_4LCl^+$ complex. Such an overlap is not possible with the corresponding *trans* complex (*c*) unless (*d*) the $d_{x^2-y^2}$ orbital is freed by rearrangement to a trigonal bipyramid intermediate. [Reproduced, with permission, from F. BASOLO and R. G. PEARSON, *Prog. Inorg. Chem.*, vol. 4, p. 448, Interscience, 1962.]

accordingly. These complexes having L groups (OH^-, N_3^-, Cl^- and SCN^-) which π-bond with donation of an electron pair to the central metal atom are believed to substitute by an S_N1 mechanism. Thus, in the hydrolysis of the *cis* isomers, the L group, through its ability to π-bond to the metal, is thought to both weaken the adjacent Co—X bond and stabilize the resultant five-coordinate intermediate, as in the conjugate-base mechanism for base hydrolysis. Figure 30 illustrates such an overlapping and donation of electrons from an orbital on the ligand L, (*a*) to a vacant *p*-orbital, or (*b*) to a vacant d^2sp^3 hybrid orbital on the metal. Similar orbital overlaps are not possible in the hydrolysis of the *trans* isomers, (*c*) unless there is a geometric rearrangement. If a trigonal bipyramid intermediate is formed, then the $d_{x^2-y^2}$ orbital can π-bond with orbitals on L, as shown in (*d*), thus stabilizing the five-coordinate intermediate and reducing activation energy requirements. Such a rearrangement will also account for the subsequent formation of a mixture of *cis* and *trans* isomers in these reactions.

When the ligand L is non-π-bonding, e.g. it is a water or ammonia molecule, or a π-electron acceptor, e.g. NO_2^-, then an S_N2 mechanism is favoured. With NO_2^-, for example, there is a movement of electrons away from the cobalt atom,

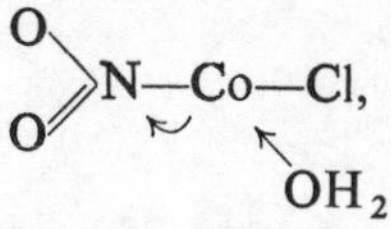

$$(O{=})(O)\rangle N—Co—Cl, \quad \leftarrow OH_2$$

the net result of which is to leave the *trans* position more open to nucleophilic attack. A similar electron displacement is thought to be possible with L = N_3^-, and the latter can, therefore, function in both ways. Thus the azide ligand can either accept or donate a pair of electrons to the cobalt atom,

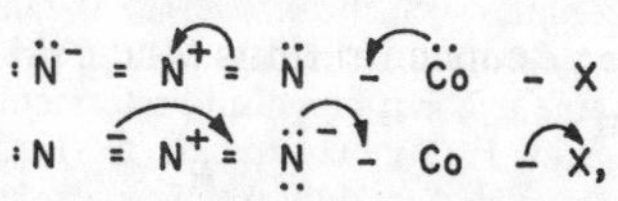

$$:\ddot{N}^- = N^+ = \ddot{N} - \ddot{Co} - X$$

$$:N \equiv N^+ - \ddot{N}^- - Co - X,$$

and, in this way, it is possible to explain the similarity in rate constants for the hydrolysis of such *cis* and *trans* forms. It follows that substitution is probably not exclusively by an S_N1 mechanism, hence the reason for there being only 20% steric change in the hydrolysis of *trans*-$Co(en)_2N_3Cl^+$.

Acid hydrolysis of bidentate ligands

In the hydrolysis of $Fe(bipy)_3^{2+}$,

$$\text{Rate} = k_{obs}[Fe(bipy)_3^{2+}],$$

k_{obs} is acid dependent, but, with increasing acid, a limiting value is reached (Fig. 31). The form of this curve can be accounted for by a mechanism

2+ (bipy)$_2$Fe(N N) $\underset{k_{-1}}{\overset{k_1}{\rightleftharpoons}}$ 2+ (bipy)$_2$Fe–N N $\xrightarrow{k_2}$ (bipy)$_2$Fe^{2+} + N N

$+H^+$ ↙ k_3

3+ (bipy)$_2$Fe–N NH $\xrightarrow{\text{fast}}$ (bipy)$_2$Fe^{2+} + + N NH

in which a small concentration of a partly dissociated species is first formed.[10] Applying stationary-state kinetics, the first-order rate constant k_{obs} can be expressed

$$k_{obs} = \frac{k_1(k_2 + k_3[H^+])}{k_{-1} + k_2 + k_3[H^+]}.$$

At low hydrogen-ion concentrations when $[H^+] \rightarrow 0$, this equation takes the form

$$k_{obs} = \frac{k_1 k_2}{k_{-1} + k_2},$$

but at high acid concentration such that $k_3[H^+] \gg k_{-1} + k_2$,

10. F. Basolo, J. C. Hayes and H. M. Neumann, *J. Am. Chem. Soc.* **75**, 5102 (1953).

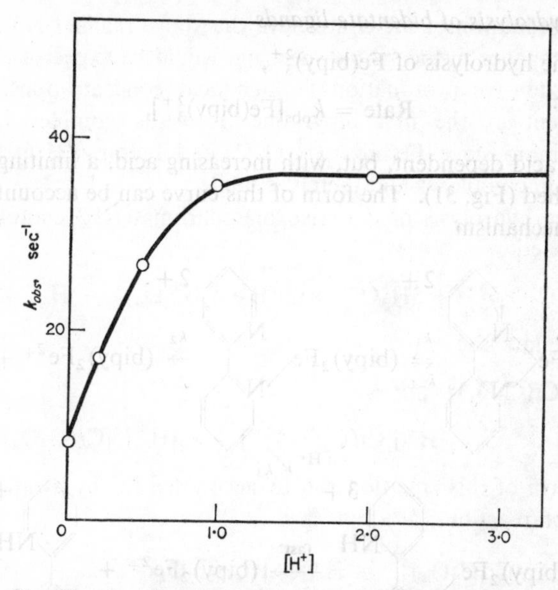

FIG. 31. Showing the effect of hydrogen-ion concentration on the experimental rate constant k_{obs} in the hydrolysis of $Fe(bipy)_3{}^{2+}$. [Reproduced, with permission, from J. H. BAXENDALE and P. GEORGE, *Trans. Faraday Soc.* **46,** 736 (1950).]

k_{obs} becomes equal to k_1. The ratio of these limiting values can be obtained from Fig. 31, i.e.

$$k_2/(k_{-1} + k_2) = 0{\cdot}16,$$

thus, on 84% of the occasions, when the first bond breaks it reforms again, while, on the other 16%, dissociation takes place.

An alternative explanation is possible, however. Thus the complex might be extensively protonated (type of bonding unspecified) but not, in this case, to a partly dissociated intermediate. If then, the protonated form hydrolyses more rapidly than the normal form, and if protonation is practically complete in 1 N acid, so that further quantities of acid can have little effect on the rate, the general shape of Fig. 31 can be explained. The first of these mechanisms is generally preferred, since evidence for extensive protonation of the complex in the above manner is lacking. Thus, changes in the visible spectrum would be expected if there were extensive protonation at high acid concentrations. Replacement of the first bipyridine from the complex is rate determining since $Fe(bipy)_2(H_2O)_2^{2+}$ and $Fe(bipy)(H_2O)_4^{2+}$ are high-spin complexes and therefore labile.

In the hydrolysis of the trisoxalatochromate(III) complex,[11] total reaction

$$Cr(C_2O_4)_3^{3-} + 3H_3O^- \rightarrow Cr(H_2O)_2(C_2O_4)_2^- + H_2C_2O_4,$$

the rate law is

$$-d[Cr(C_2O_4)_3^{3-}]/dt =$$
$$k_1[H^+][Cr(C_2O_4^-)_3^{3-}] + k_2[H^+]^2[Cr(C_2O_4)_3^{3-}].$$

The form of this equation can be accounted for by a rapid protonation reaction,

$$Cr(C_2O_4)_3^{3-} + H^+ \overset{\text{fast}}{\rightleftharpoons} HCr(C_2O_4)_3^{2-},$$

which is followed by parallel rate determining steps

$$HCr(C_2O_4)_3^{2-} + 2H_2O \rightarrow Cr(C_2O_4)_2(H_2O)_2^- + HC_2O_4^-,$$
$$HCr(C_2O_4)_3^{2-} + H^+ + 2H_2O \rightarrow Cr(C_2O_4)_2(H_2O)_2^- + H_2C_2O_4.$$

At 50°C the rate constants, as defined above, were found to be $k_1 = 1{\cdot}7 \times 10^{-4}$ l mole^{-1} sec^{-1} and $k_2 = 5{\cdot}0 \times 10^{-4}$ l^2 mole^{-2} sec^{-1}.

Hydrolysis of ethylenediamine complexes of the type $Ni(en)_3^{2+}$ and $Cr(en)_3^{3+}$ are also acid dependent, the rate of replacement slowing down as successive ligand groups are removed. Further details of the dissociation of ethylenediamine from the complex $Ni(en)_3^{2+}$ will be given in the next section. Not all reactions involving the hydrolysis of bidentate ligands have hydrogen-ion dependent paths however. Thus there is no pH dependence in the hydrolysis of the tris-*o*-phenanthroline complexes $Fe(phen)_3^{2+}$,

---

11. K. V. Krishnamurty and G. M. Harris, *J. Phys. Chem.* **64**, 346 (1960).

$Co(phen)_3^{2+}$, and $Ni(phen)_3^{2+}$. This is probably due to the rigidity of the *o*-phenanthroline ligand. Contrast the behaviour of the intermediate formed in the hydrolysis of $Fe(bipy)_3^{2+}$,

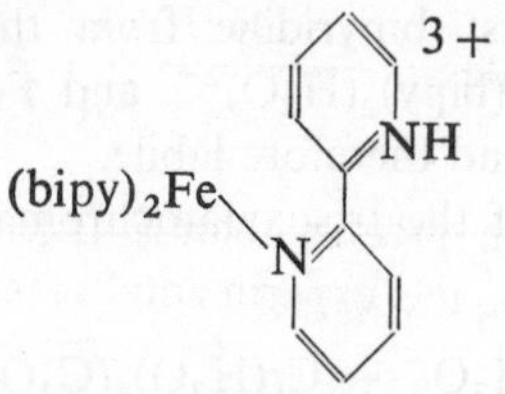

$$(bipy)_2Fe-N(C_5H_4)-(C_5H_4)NH^{3+}$$

which can readily become protonated at the uncoordinated nitrogen atom (as shown), since there is free rotation of this ring.

The hydrolysis of some nickel(II) amine complexes

The rates of dissociation of Ni^{II} complexes with a series of nitrogen-containing ligands have been studied spectrally using, in most cases, the stopped-flow method.(12)

In the dissociation of the mono complexes of monodentate ligands,

$$NiL^{2+} \rightarrow Ni^{2+} + L,$$

reactions are found to be independent of the hydrogen-ion concentration. At 25°C and in 0·2 N nitric acid solutions, the rate constants for the dissociation of ammonia, pyridine, and hydrazine, are 5·8, 5·0 and 3·6 sec^{-1}, respectively. The corresponding rates of formation which are obtained from known stability constants (the dissociation is much simpler to investigate than the formation), are very similar to one another, and, at 25°C, lie within the range 2·5–3·2 × 10^3 l $mole^{-1}$ sec^{-1}. Because of this an S_N1 mechanism seems likely,

$$Ni(H_2O)_6^{2+} \underset{k_{-1}}{\overset{k_1}{\rightleftharpoons}} Ni(H_2O)_5^{2+} + H_2O$$

$$Ni(H_2O)_5^{2+} + L \xrightarrow{k_2} Ni(H_2O)_5L^{2+},$$

---

12. AHMED A. K. SHAMSUDDIN and R. G. WILKINS, *J. Chem. Soc.* 3700 (1959) and 2895, 2901 (1960).

and, assuming stationary-state kinetics for $Ni(H_2O)_5^{2+}$, the rate of formation of $Ni(H_2O)_5L^{2+}$ is given by the expression

$$\text{Rate} = \frac{k_1 k_2 [Ni(H_2O)_6^{2+}][L]}{k_{-1} + k_2[L]}.$$

If k_{-1} is very much greater than $k_2[L]$, a second-order rate equation is obtained, the experimental rate constant k_{obs} being equal to $k_1 k_2 / k_{-1}$.

Conventional methods have been used to study the dissociation of the mono(ethylenediamine)nickel ion, and related bidentate complexes. The dissociation may be expressed as follows

$$Ni^{2+}\left\langle \begin{array}{l} NH_2 \\ | \\ CH_2 \\ | \\ CH_2 \\ | \\ NH_2 \end{array} \right. \underset{k_{-3}}{\overset{k_3}{\rightleftharpoons}} Ni^{2+} \diagup \begin{array}{l} NH_2 \\ | \\ CH_2 \\ | \\ CH_2 \\ | \\ NH_2 \end{array} \underset{k_{-4}}{\overset{k_4}{\rightleftharpoons}} Ni^{2+} + \begin{array}{l} NH_2 \\ | \\ CH_2 \\ | \\ CH_2 \\ | \\ NH_2 \end{array}$$

Hydrogen ions are believed to enhance the dissociation in two ways, at lower acidities by altering the form of the intermediate (i.e. $Ni{-}NH_2{-}CH_2{-}CH_2{-}NH_3^{3+}$ is formed), and at high acidities when the hydrated proton can assist more directly in the rupture of the first Ni—N bond. Over the pH range 1·5–4·0, the reaction is independent of hydrogen-ion concentration, and under these conditions the observed rate constant can be equated to k_3, since k_3 must be slow and k_4 relatively fast. The dissociation of second and third ligand groups has been studied in a similar manner. Results for ethylenediamine and butylenediamine (2,3-diaminobutane) are shown in Table 32.

More recently, the step-wise nature of the dissociation of Ni^{II}–polyamine complexes has been studied using the stopped-flow method.[13] The decomposition was examined in an excess

13. G. A. Melson and R. G. Wilkins, *J. Chem. Soc.* 2662 (1963).

of acid, there being little dependence on the hydrogen-ion concentration once this is greater than 0·05 M. The proton acts as an effective scavenger for the released amine group without aiding bond rupture. In the reaction with terdentate ligands, the first two stages are rate determining. Thus, with triaminopropane, the reaction I → II can, in the first instance, be measured, the subsequent decomposition (II) → (III) being much slower,

$$\begin{array}{l} CH_2\,.\,NH_2 \\ |\\ CH\,.\,NH \\ | \\ CH_2\,.\,NH_2 \end{array} \!\!> Ni^{2+} \xrightarrow{H^+} \begin{array}{l} CH_2\,.\,NH_3^+ \\ | \\ CH\,.\,NH_2 \\ | \\ CH_2\,.\,NH_2 \end{array} \!\!> Ni^{2+} \xrightarrow{H^+}$$

(I) (II)

$$\begin{array}{l} CH_2\,.\,NH_3^+ \\ | \\ CH\,.\,NH_3^+ \\ | \\ CH_2\,.\,NH_2 \end{array} \!\!/ Ni^{2+} \xrightarrow[fast]{H^+} \begin{array}{l} CH_2\,.\,NH_3^+ \\ | \\ CH\,.\,NH_3^+ \\ | \\ CH_2\,.\,NH_3^+ \end{array} + Ni^{2+}.$$

(III) (IV)

There are two observable stages in the dissociation of the quadridentate tri-(2-aminoethyl)amine ligand, $N\,.\,(C_2H_5NH_2)_3$,

TABLE 32

KINETIC DATA FOR THE FIRST-ORDER DISSOCIATION OF NICKEL–DIAMINE COMPLEXES IN 0·2 M ACID AT 25°C

Complex	Rate constant (sec^{-1})	$\Delta H^\ddagger$ (kcal $mole^{-1}$)
$Ni(en)_3{}^{2+}$	86·6	18·0
$Ni(en)_2{}^{2+}$	5·2	19·9
$Ni(en)^{2+}$	0·145	20·5
$Ni(bn)_3{}^{2+}$	8·25	14·6
$Ni(bn)_2{}^{2+}$	0·257	16·1
$Ni(bn)^{2+}$	0·020	18·4

which can be written in the abbreviated form tren. These correspond to the formation of mono- and di-protonated species, respectively,

$$\mathrm{Ni(tren)^{2+} \xrightarrow{H^+} Ni(H\ tren)^{3+}},$$

$$\mathrm{Ni(H\ tren)^{3+} \xrightarrow{H^+} Ni(H_2\ tren)^{4+}},$$

$$\mathrm{Ni(H_2\ tren) \xrightarrow[fast]{H^+} Ni^{2+} + H_3\ tren^{3+}}.$$

Examination of models shows that the tertiary nitrogen atom cannot be the first group removed, and the first stage is, therefore,

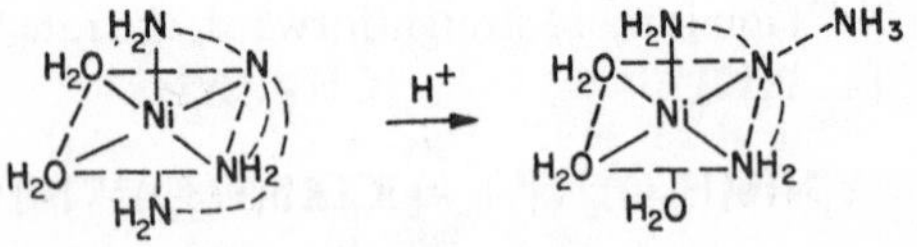

The second stage probably involves the breakage of a bond to one of the other coordinated primary amine groups.

The energies of activation for the first bond rupture in all these cases fall neatly into groups. It is highest for the bidentate ethylendiamine complex, containing the most strain-free ring ($\sim$21 kcal mole^{-1}), around 18 kcal mole^{-1} for complexes of terdentate ligands, and only 15 kcal mole^{-1} for those complexes with quadri- and quinque-dentate amines, and with ammonia itself.

The substitution of some 3+ *aquo ions*

The slow exchange of water molecules between the aquochromium(III) (t_{2g}^3) ion and solvent, $k = 3{\cdot}2 \times 10^{-6}$ sec^{-1} at 27°C, has been measured using oxygen-18 labelled solvent,

$$\mathrm{Cr(H_2O)_6^{3+} + H_2{}^{18}O \rightarrow Cr(H_2O)_5(H_2{}^{18}O)^{3+} + H_2O}.$$

Such experiments also serve to confirm the hydration number of six since it can be demonstrated that six water molecules are held back per metal ion.

In the case of the hexaquo Co^{3+} (t_{2g}^6) ion, exchange of water molecules appears to proceed more readily than might at first be expected. This may be due to the existence of small equilibrium concentrations of the spin-free $t_{2g}^4e_g^2$ form of Co^{3+}, as has previously been indicated, substitution of the latter being more rapid than that of the t_{2g}^6 spin-paired form. In addition, the Co^{3+} hexaquo ion is known to oxidize water so that solutions are never completely free from labile Co^{2+}. As a result of this, a complete substitution of the inner sphere of the Co^{3+} ion can proceed by electron transfer between the Co^{2+} and Co^{3+} ions.

The slow exchange of water between solvent and hexaquo-rhodium(III) ($4d^6$) ion is more straightforward, the rate law taking the form

$$\text{Rate} = k_1[\text{Rh(H}_2\text{O)}_6^{3+}] + k_2K_a[\text{Rh(H}_2\text{O)}_6^{3+}]/[\text{H}^+]$$

where k_1 and k_2 are first-order rate constants for the replacement of an H_2O molecule from the hexaquo and pentaquohydroxo ions respectively, and K_a is the first acid dissociation constant of the hexaquorhodium(III) cation.(14) At 64·4°C and $\mu = 12$ M the magnitudes of the three constants are $2{\cdot}3 \times 10^{-5}$ sec^{-1}, $3{\cdot}3 \times 10^{-3}$ sec^{-1} and $6{\cdot}6 \times 10^{-5}$ mole l^{-1} respectively, and the activation energy corresponding to k_1 is 33 kcal mole^{-1}. In these experiments, high ionic strengths were used, since the concentration of Rh^{3+} was in the range of 0·6–1·7 M.

Again, by using isotopic dilution techniques, the hydration number of the Al^{3+} ion, has been shown to be six. In the exchange of $Al(H_2O)_6^{3+}$ with H_2O at 25°C, the half-life is > ·02 sec, and Al^{3+} is more inert than Ni^{2+} and Fe^{3+} (it is a smaller ion).(15) Fast reaction techniques have been used for more labile 3+ metal ions, and a rate constant of $2{\cdot}4 \times 10^4$ sec^{-1} at 26°C has been

14. W. Plumb and G. M. Harris, *Inorg. Chem.* 3, 542 (1964).
15. H. W. Baldwin and H. Taube, *J. Chem. Phys.* 33, 206 (1960).

obtained for the substitution of the spin-free hexaquo Fe^{3+} ($t_{2g}^3 e_g^2$) ion using NMR techniques. Much higher rate constants in the range 10^6–10^8 sec^{-1}, which have been obtained for the rare-earth metal ions Sc^{3+}, Y^{3+} and La^{3+}, suggest that these ions have co-ordination numbers greater than six.[16]

Substitution reactions of the Fe^{3+} hexaquo ion,

$$Fe^{III}(H_2O)_6^{3+} + Y^- \rightarrow Fe^{III}(H_2O)_5Y^{2+} + H_2O,$$

have been fairly extensively studied using rapid flow techniques.[17] Their rates may be expressed

$$-d[Fe^{3+}]/dt = k_1[Fe^{3+}][Y^-] + k_2[Fe^{3+}][Y^-]/[H^+]$$

where $Y^- = Cl^-$, Br^- and NCS^-, and k_2 corresponds to the substitution of the hydrolysed ion $Fe(H_2O)_5OH^{2+}$ (Table 33). With fluoride the rate law is of the form

$$-d[Fe^{3+}]/dt = k_a[Fe^{3+}][F^-][H^+] + k_b[Fe^{3+}][F^-].$$

Alternatively this may be expressed

$$-d[Fe^{3+}]/dt = k_1[Fe^{3+}][HF] + \frac{k_2[Fe^{3+}][HF]}{[H^+]}$$

which is of the same form as the above with $Y^- = HF$. Azide shows the same sort of behaviour. For k_2 an S_N1 CB mechanism

TABLE 33
KINETIC DATA FOR THE FORMATION OF COMPLEXES OF Fe^{3+} + AT 25°C

Y^-	k_1 ($l\ mole^{-1}\ sec^{-1}$)	k_2 (sec^{-1})
Cl^-	9·4	18
Br^-	20	31
NCS^-	127	20
HN_3	2·6	10
HF	4·0	11·2

16. M. Eigen, personal communication.
17. G. G. Davis and W. MacF. Smith, *Canad. J. Chem.* **40**, 1836 (1962).

seems likely since observed rate constants are numerically so very similar. Thus, the OH^- group through its ability to π-bond to the central metal atom, helps to free an adjacent water molecule in an S_N1 dissociation process,

$$Fe(H_2O)_5OH^{2+} \underset{k_r}{\overset{k_d}{\rightleftharpoons}} Fe(H_2O)_4OH^{2+} + H_2O$$

$$Fe(H_2O)_4OH^{2+} + Y^- \xrightarrow{k_a} Fe(H_2O)_4OHY^+.$$

Assuming stationary-state kinetics for $Fe(H_2O)_4OH^{2+}$, the rate of substitution is given by an expression

$$\text{Rate} = \frac{k_d k_a [Fe(H_2O)_5OH^{2+}][Y^-]}{k_r + k_a[Y^-]},$$

and if $k_r \gg k_a[Y^-]$, it follows that

$$\text{Rate} = \frac{k_d k_a}{k_r} [Fe(H_2O)_5OH^{2+}][Y^-].$$

Making a further substitution for $[Fe(H_2O)_5OH^{2+}]$

$$\text{Rate} = \frac{k_d k_a K_h}{k_r} \frac{[Fe^{3+}][Y^-]}{[H^+]},$$

where $K_h \sim 10^{-3}$ l^{-1} mole is for the equilibrium

$$Fe(H_2O)_6^{3+} \rightleftharpoons Fe(H_2O)OH^{2+} + H^+.$$

The rate constant k_2 can, therefore, be identified as $k_d k_a K_h / k_r$, and since K_h is $1{\cdot}65 \times 10^{-3}$ l mole^{-1} at 25°, the composite rate constant $k_d k_a / k_r$ for the substitution of $Fe(H_2O)_5OH^{2+}$ is $\sim 10^3$ times k_1 for the substitution of Fe $(H_2O)_6^{3+}$. The small spread in k_1 values, Table 33, is also consistent with an S_N1 mechanism. Prior formation of an outer-sphere complex is possible; an S_N2 mechanism seems unlikely in view of the k_2 hydroxo activation.

The substitution of some other labile aquo ions

The simplest and certainly the most fundamental reactions to consider here are those in which a coordinated water molecule is exchanged with a solvent water molecule. Such reactions can be studied by NMR methods,[18] but by measuring the proton resonance alone, it is never quite certain whether protons move independently of the oxygen or whether the whole H_2O molecule is exchanged. Of the three oxygen isotopes, oxygen-17 alone has a nuclear spin and by using water containing $H_2{}^{17}O$ information as to the movement of oxygen atoms can be obtained. Rate constants for water exchange reactions with 2+ transition metal ions are shown in Table 34.

TABLE 34
RATE CONSTANTS FOR $M(H_2O)_6{}^{2+} + H_2O \rightarrow$, EXCHANGE REACTIONS AT ROOM TEMPERATURE AS DETERMINED BY THE NMR METHOD

Metal ion	Rate constant (sec^{-1})	Metal ion	Rate constant (sec^{-1})
Ni^{2+}	$2{\cdot}7 \times 10^4$	Mn^{2+}	$3{\cdot}1 \times 10^7$
Co^{2+}	$1{\cdot}1 \times 10^6$	Cr^{2+}	$7{\cdot}4 \times 10^9$
Fe^{2+}	$3{\cdot}2 \times 10^6$	Cu^{2+}	$8{\cdot}3 \times 10^9$

Eigen has used relaxation methods to follow the rate of conversion of outer-sphere complexes (with different ligands ranging from weakly complexing species, such as the nitrate ion, to strongly chelating ligands such as EDTA) to inner-sphere complexes, and in this way has been able to indicate characteristic rate constants for the substitution of inner-sphere water molecules, Fig. 32. For the 2+ transition metal ions these rates are of the same order of magnitude as those listed in Table 34, and a similar correspondence is expected to hold for other ions. The relative rates for the reaction of doubly charged transition ions, Fig. 32, can be accounted for by considering ligand-field effects. Thus the loss of ligand-field stabilization energy for the $d^3(t_{2g}^3)$ and $d^8(t_{2g}^6e_g^2)$ configurations is particularly high whether S_N1 or S_N2 mechanisms are involved, and the substitution of V^{2+} and high-spin Ni^{2+} are therefore expected to be slower than for other doubly chaıged cations. This has been amply verified in the case of

18. R. E. CONNICK and T. J. SWIFT, *J. Chem. Phys.* **37**, 301 (1962).

Ni^{2+}, relatively little work has been done with V^{2+} owing to the ease with which it is oxidized. The high rate of substitution for $Cu^{2+}(t_{2g}^6 e_g^3)$ is attributed to the distorted octahedral structure of this aquo-ion owing to the unsymmetrical electronic structure. In consequence, two axial bonds are much longer than the other four and substitution of the weakly held axial ligands is very rapid. For the alkali and alkaline-earth metal ions solvent interactions are much weaker, and coordination numbers less clearly defined than with transition metal ions. Rate constants for the

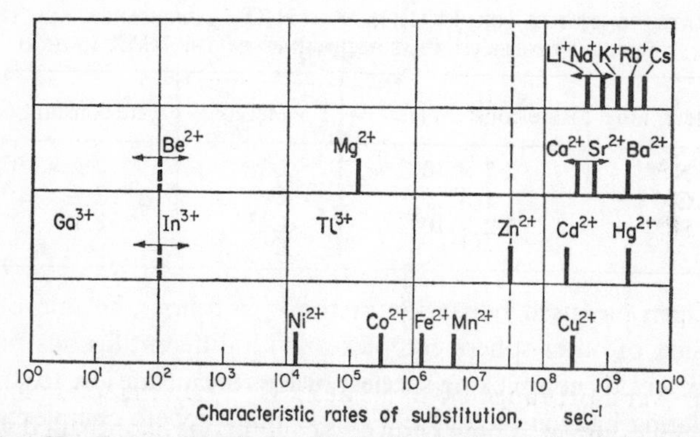

FIG. 32. Rate constants for the H_2O substitution in the inner coordination sphere of metal ions. [Reproduced, with permission, from M. EIGEN, *Seventh Conference on Coordination Chemistry*, p. 105, Butterworth, London, 1963.]

substitution of the inner sphere of solvent molecules around the ions $Li^+ < Na^+ < K^+ < Rb^+ < Cs^+$ are in the order of ionic radii, as would be expected, while the much slower rates for the aquo Mg^{2+} and Be^{2+} ions are in keeping with the increased tendency of these ions to form bonds of a covalent nature.

Substitution of the central metal atom

In these reactions, the central metal atom of a complex is replaced by a second metal ion which is initially present as an aquo ion. The easiest systems to study are those involving a potentially hexadentate ligand such as ethylenediaminetetra-acetate, which for present purposes may be written Y^{4-}. For systems in which the incoming metal ion is of the same kind as that being replaced, one of the metal ions has to be labelled. The exchange between Ni^{2+} and the nickel(II) chelate of ethylenediaminetetra-acetate has, for example, been studied in some detail over a pH range 1·0 to 5·3 at 25°C.[19] In the solid phase at least this complex is now known to contain one coordinated water molecule, with the chelate attached at five and not six positions (through two nitrogen and three oxygen atoms), and the same is probably true for the complex in solution. The reaction has a complex rate equation[20] which may be written

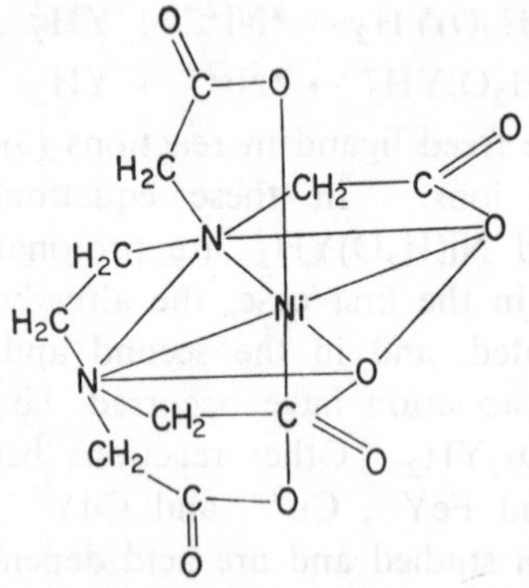

$$\text{Rate} = k_1[\text{Ni}(\text{H}_2\text{O})\text{Y}^{2-}][\text{Ni}^{2+}] + k_2[\text{H}^+][\text{Ni}(\text{H}_2\text{O})\text{Y}^{2-}][\text{Ni}^{2+}] + k_3[\text{H}^+][\text{Ni}(\text{H}_2\text{O})\text{Y}^{2-}] + k_4[\text{H}^+]^2[\text{Ni}(\text{H}_2\text{O})\text{Y}^{2-}] + k_5[\text{H}^+]^3[\text{Ni}(\text{H}_2\text{O})\text{Y}^{2-}]$$

19. G. S. Smith and J. L. Hoard, *J. Am. Chem. Soc.* **81**, 556 (1959).
20. C. M. Cook and F. A. Long, *J. Am. Chem. Soc.* **80**, 33 (1958).

where rate constants k_1 and k_2 probably correspond to exchange paths

$$*Ni(H_2O)Y^{2-} + Ni^{2+} \rightarrow Ni(H_2O)Y^{2-} + *Ni^{2+}, \quad (1)$$

$$*Ni(H_2O)HY^{-} + Ni^{2+} \rightarrow Ni(H_2O)Y^{2-} + *Ni^{2+} + H^{+}, \quad (2)$$

and k_3, k_4 and k_5 indicate initial (rate determining) dissociation reactions of protonated $Ni(H_2O)Y^{2-}$ species, thus

$$*Ni(H_2O)YH^{-} \rightarrow *Ni^{2+} + YH^{3-}, \quad (3)$$

$$*Ni(H_2O)YH_2 \rightarrow *Ni^{2+} + YH_2^{2-}, \quad (4)$$

$$*Ni(H_2O)YH_3^{+} \rightarrow *Ni^{2+} + YH_2. \quad (5)$$

Subsequently, the freed ligand in reactions (3)–(5) re-complexes with free Ni^{2+} ions. In these equations $Ni(H_2O)YH^{-}$, $Ni(H_2O)YH_2$ and $Ni(H_2O)YH_3^{+}$ are protonated forms of the complex, where, in the first case, the already free carboxylate group is protonated, and in the second and third cases dissociation and protonation have occurred, i.e. $Ni(H_2O)YH_2$ is probably $Ni(H_2O)_2YH_2$. Other reactions between Co^{2+} and CoY^{2-}, Fe^{3+} and FeY^{-}, Cu^{2+} and CdY^{2-}, and Zn^{2+} and PbY^{2-} have been studied and are acid dependent in a similar manner.

An interesting variation on this type of reaction is to study the effect which β-decay of the central metal atom has on the rate of exchange of the central metal atom. As might be expected, the daughter atom, because of the recoil action, escapes much more readily from its chelate environment than it normally would. Consider, for example, the ethylenediaminetetra-acetate complexes of Yb^{3+}, which is normally very resistant to thermal exchange in the presence of its free metal ions. When the unstable $^{177}Yb^{3+}$ isotope is used (half-life 1·8 hr), some 56% of the $^{177}Lu^{3+}$ ion which is formed by β-decay escapes from the complex environment at the time of decay.[21] By having an excess of free Yb^{3+} to complex with the Y^{4-} ligand as it is set free, the $^{177}LuY^{-}$ complex has little opportunity of reforming, and the net chemical change can be followed.

21. P. Glentworth and R. H. Betts, *Canad. J. Chem.* **39**, 1049 (1961).

CHAPTER 13

SUBSTITUTION REACTIONS OF SQUARE-PLANAR COMPLEXES

SQUARE-planar complexes are most common for metal ions having d^8 configuration, e.g. Pt^{II}, Pd^{II}, Ni^{II} and Au^{III}. Of these, substitution reactions of Pt^{II} complexes has been extensively studied, since these complexes are most stable and are relatively inert to substitution. This chapter is primarily concerned with the reactions of platinum(II); it is to be expected that other square-planar systems will show similar behaviour.

The difference between square-planar and octahedral complexes is not always clearcut, since square-planar complexes are known to form weak bonds with solvent molecules above and below the plane of the complex, and to this extent are six-coordinated. Magnetic susceptibility measurements are particularly useful in allowing a distinction to be made between such forms. Thus Ni^{II} forms a wide range of octahedral complexes which are spin-free and paramagnetic, e.g. $Ni(H_2O)_6^{2+}$, $Ni(NH_3)_6^{2+}$, $Ni(en)_3^{2+}$, while other Ni^{II} and all Pd^{II} and Pt^{II} complexes are diamagnetic and essentially square planar. Note, incidentally, that the simple aquo ion of Pt^{II} is unknown.

Preparation of cis *and* trans *isomers: the* trans-*effect*

Two isomeric forms of the complex $Pt^{II}(NH_3)_2Cl_2$ are known, and since only one tetrahedral arrangement of these ligands is possible, and both complexes are diamagnetic, clearly these must both have square-planar configuration. The *trans* complex can be prepared by heating the chloric salt of $Pt(NH_3)_4^{2+}$ to $\sim 250°C$,

$$\begin{matrix} NH_3 & & NH_3 \\ & Pt & \\ NH_3 & & NH_3 \end{matrix} \xrightleftharpoons[-NH_3]{+Cl^-} \begin{matrix} NH_3 & & Cl \\ & Pt & \\ NH_3 & & NH_3 \end{matrix} \xrightleftharpoons[-NH_3]{+Cl^-} \begin{matrix} NH_3 & & Cl \\ & Pt & \\ Cl & & NH_3 \end{matrix}$$

and the *cis* form by heating $PtCl_4^{2-}$ in buffered ammonia solution

$$\begin{matrix} Cl & & Cl \\ & Pt & \\ Cl & & Cl \end{matrix} \xrightleftharpoons[-Cl^-]{+NH_3} \begin{matrix} Cl & & NH_3 \\ & Pt & \\ Cl & & Cl \end{matrix} \xrightleftharpoons[-Cl^-]{+NH_3} \begin{matrix} Cl & & NH_3 \\ & Pt & \\ Cl & & NH_3 \end{matrix}$$

The second stage in each of these is of interest because a preferential replacement of groups *trans* to Cl^- is indicated. By reversing the order in which $PtCl_4^{2-}$ is treated with NH_3 and NO_2^-, the two isomeric forms of $Pt(Cl)_2(NH_3)NO_2^-$ can be prepared,

$$\begin{matrix} Cl & & Cl \\ & Pt & \\ Cl & & Cl \end{matrix} \xrightleftharpoons[-Cl^-]{+NH_3} \begin{matrix} Cl & & NH_3 \\ & Pt & \\ Cl & & Cl \end{matrix} \xrightleftharpoons[-Cl^-]{+NO_2^-} \begin{matrix} Cl & & NH_3 \\ & Pt & \\ Cl & & NO_2 \end{matrix}$$

$$\begin{matrix} Cl & & Cl \\ & Pt & \\ Cl & & Cl \end{matrix} \xrightleftharpoons[-Cl^-]{+NO_2^-} \begin{matrix} Cl & & NO_2 \\ & Pt & \\ Cl & & Cl \end{matrix} \xrightleftharpoons[+Cl^-]{+NH_3} \begin{matrix} Cl & & NO_2 \\ & Pt & \\ NH_3 & & Cl \end{matrix}$$

The dependence of the rate of substitution of a ligand upon the nature of the group in the *trans* position is known as the *trans*-effect. For the above reactions, the order of effectiveness is $NO_2^- > Cl^- > NH_3$ and from other similar studies this series can be extended to read CN^-, CO, C_2H_4, $NO < PR_3$, $SR_2 > NO_2^- > I^-$, $SCN^- > Br^- > Cl^- > NH_3$, py, $RNH_2 > OH^- > H_2O$, where R is an alkyl or aryl group. It should be emphasized that the *trans*-effect indicates only the predominant and not the exclusive formation of an isomer.

A theoretical account of the trans-*effect*

There are two main theoretical approaches to consider in accounting for the *trans*-effect. In the first, the *trans*-directing ligand L in a complex $Pt^{II}ABLX$ is believed to weaken the bond to the *trans* group X by electrostatic effects. Thus, the residual charge on the central metal atom first induces a dipole on the *trans*-directing ligand which, in turn, induces a dipole on the central metal atom, as shown in Fig. 33. The net effect of such an induced dipole is to weaken the Pt—X bond as shown. The theory depends on the mutual polarizability of the central metal atom and L, where the polarizability of L is assumed to be greater than that of any other ligand present.

With some ligands, the halides, in particular, the correlation between polarizability and the *trans*-effect is good. For others, however, the ability to π-bond by accepting electrons from the central-metal atom appears to be more important and CO and C_2H_4 at least are believed to function as strong *trans* activators in this way. Thus, if there is π-bonding between a filled *d*-orbital on the metal with a vacant *p*-orbital on L, as shown in Fig. 34b, there is a shift in electron density away from the X group. The net effect of the π-bonding is not to weaken the Pt–X bond, but to reduce the electron density in the vicinity of X and, in this way, facilitate approach of the incoming ligand Y.

The effect of varying the incoming ligand

The rate equation for the substitution of the ligand X from a complex $Pt^{II}ABLX$ consists of two terms, one first and the other second order,

$$\frac{-d[Pt^{II}ABLX]}{dt} = k_1[Pt^{II}ABLX] + k_2[Pt^{II}ABLX][Y],$$

where Y is the incoming ligand. Details of mechanisms which will account for both these paths will be considered in a later section.

The effect which different Y groups have on the rate has been studied for a series of reactions,

$$\text{Pt(dien)Cl}^+ + \text{Y} \rightarrow \text{Pt(dien)Y}^{2+} + \text{Cl}^-,$$

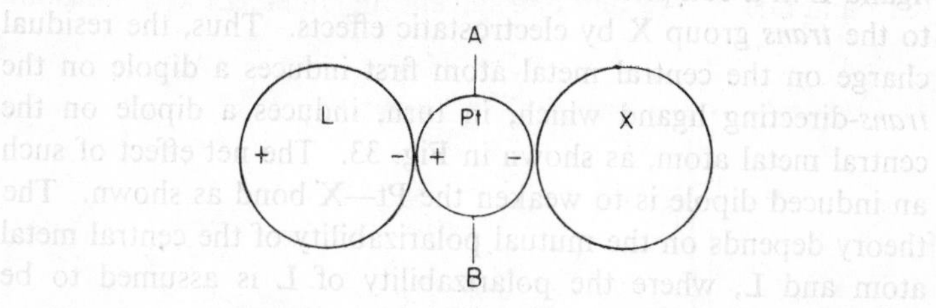

FIG. 33. Induction effects between the central metal atom and the ligand L of a complex $Pt^{II}ABLX$ can result in a weakening of the Pt—X bond.

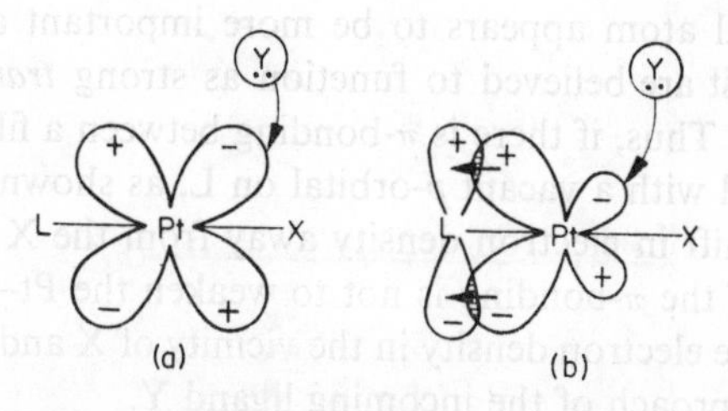

FIG. 34. The *trans*-effect can result from π-bonding of the ligand L (acceptor) to the central metal atom. In (*a*) L does not form π-bonds, but in (*b*) there is π-bonding with a reduction in electron density in the vicinity of X.

where dien is the tridentate ligand diethylenetriamine ($NH_2 . CH_2 . CH_2 . NH . CH_2 . CH_2 . NH_2$).[1] Pseudo first-order rate constants k_{obs} are obtained by having Y in large excess so that

$$-d[\text{Pt(dien)Cl}^+]/dt = k_{obs}[\text{Pt(dien)Cl}^+],$$

1. D. BANERJEA, F. BASOLO and R. G. PEARSON, *J. Am. Chem. Soc.* **79**, 4055 (1957); H. B. GRAY, *ibid.* **84**, 1548 (1962).

where k_{obs} shows the following linear dependence on [Y],

$$k_{obs} = k_1 + k_2[Y].$$

In Fig. 35, plots of k_{obs} against [Y] are shown for Y = I^-, Br^-, Cl^- and OH^-. For any one solvent, the intercept k_1 is constant

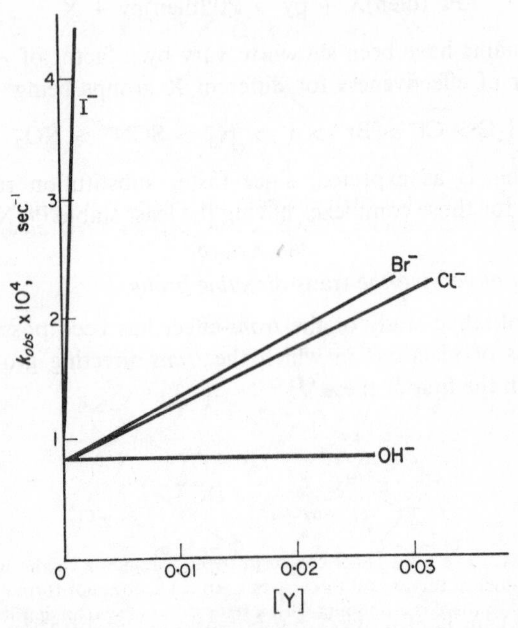

FIG. 35. Dependence of k_{obs} on the concentration of the incoming ligand Y for reactions with $Pt(dien)Cl^+$ at 25°C. [Reproduced, with permission, from H. B. GRAY, *J. Am. Chem. Soc.* **84**, 1548 (1962).]

and this path is said to be solvent controlled. The gradient k_2, on the other hand, is a measure of the reactivity of Y, thus $PR_3 \sim SCN^- \sim I^- > N_3^- \sim NO_2^- > py \sim NH_3 \sim Br^- > Cl^- > OH^- \sim H_2O$. The order here is similar to that of the *trans*-effect series which suggests that a good *trans*-labilizing group is also a good substituting ligand. This is as would be expected if reactions proceed by an S_N2 mechanism in which the activated complex contains the entering Y group.

The effect of varying the leaving group

For a series of reactions in which the leaving group X is varied,

$$Pt^{II}(dien)X + py \rightarrow Pt^{II}(dien)py + X,$$

rate constants have been shown to vary by a factor of $\sim 10^5$,[2] the order of effectiveness for different X groups being

$$NO_3^- > H_2O > Cl^- > Br^- > I^- > N_3^- > SCN^- > NO_2^- > CN^-.$$

Again, this is as expected, since faster substitution rates are observed for those complexes having the least stable Pt–X bonds.

The effect of varying the trans-*directing group*

A quantitative study of the *trans*-effect has been possible for two series of reactions in which the *trans*-directing gıoup L is varied. In the first of these,[3]

L --- NH_3 / Pt / Cl --- Cl + py → L --- NH_3 / Pt / Cl --- py + Cl^-

reactions can be followed by potentiometric titration of the unreacted pyridine. Second-order rate constants only,

$$-d[Pt^{II}Cl_2LNH_3]/dt = k_2[Pt^{II}Cl_2LNH_3][py],$$

are quoted in the original paper, and the solvent controlled path k_1 is presumably small. The *trans*-labilizing effect of different L

2. F. Basolo, H. B. Gray and R. G. Pearson, *J. Am. Chem. Soc.* **82**, 4200 (1960).
3. O. E. Zvyagintsev and E. F. Karandashova, *Dokl. Akad. Nauk., SSSR* **101,** 93 (1955).

groups at 25°C, $Cl^- : Br^- : NO_2^- : C_2H_4$, is approximately 1 : 3 : 9 : > 100 (see Table 35).

For a second series of reactions in ethanol,[4]

$$\begin{matrix} L & --- & PEt_3 \\ & Pt & \\ PEt_3 & --- & Cl \end{matrix} \; + py \longrightarrow \; \begin{matrix} L & --- & PEt_3 \\ & Pt & \\ PEt_3 & --- & py \end{matrix} \; + Cl^-,$$

rates have been obtained by measuring the change in conductivity of reaction solutions. The rate law is of the form

$$-d[Pt^{II}(PEt_3)_2LCl]/dt = k_1[Pt^{II}(PEt_3)_2LCl] + k_2[Pt^{II}(PEt_3)_2LCl][Y]$$

TABLE 35

REACTION OF $Pt^{II}(Cl)_2LNH_3$ COMPLEXES WITH PYRIDINE AT 25°C AND EFFECT OF VARYING THE *trans*-DIRECTING GROUP L

L	k_2	$\Delta H^{\ddagger}$
Cl^-	6·3	19
Br^-	18	17
NO_2^-	56	11
C_2H_4	very fast	—

where both k_1 and k_2 may vary by as much as 10^4 as L is varied (Table 36). With L = Me, C_6H_5, Cl and H, the *trans*-effect is believed to be the result of inductive effects, as previously indicated. In such complexes, evidence for the weakening of the bond to the *trans* ligand has been obtained by a variety of methods including X-ray crystallography. An X-ray study of the *trans*-$Pt(PEt_3)_2HBr$ complex, for example, has shown that the Pt—Br

4. F. BASOLO, J. CHATT, H. B. GRAY, R. G. PEARSON and B. L. SHAW, *J. Chem. Soc.* 2207 (1961).

bond distance is 2·56 Å, which is significantly greater than the expected value of 2·43 Å. In the hydrido complexes, the hydrogen ligand is believed to be anionic in character.

TABLE 36

REACTION OF $Pt^{II}(PEt_3)_2LCl$ COMPLEXES WITH PYRIDINE IN ETHYL ALCOHOL AT 25°C AND EFFECT OF VARYING THE *trans*-DIRECTING GROUP L

Complex	*trans*-Directing group	k_1 (sec^{-1})	k_2 (l mole^{-1} sec^{-1})
trans-$Pt(PEt_3)_2HCl$†	H	$1{\cdot}83 \times 10^{-2}$	4·1
cis-$Pt(PEt_3)_2Cl_2$	PEt_3	$1{\cdot}67 \times 10^{-2}$	38
trans-$Pt(PEt_3)_2MeCl$	Me	$1{\cdot}67 \times 10^{-4}$	$6{\cdot}66 \times 10^{-2}$
trans-$Pt(PEt_3)_2C_6H_5Cl$	C_6H_5	$3{\cdot}34 \times 10^{-5}$	$1{\cdot}58 \times 10^{-2}$
trans-$Pt(PEt_3)_2Cl_2$	Cl	10^{-6}	$4{\cdot}0 \times 10^{-4}$

† Rate at 0°; the reaction is too fast to measure at 25°.

The mechanism for ligand substitution of square-planar complexes

Weakly bonded solvent molecules and potential ligand groups in the fifth and sixth-coordination positions are believed to play an essential role in the substitution of square-planar complexes. Both the first- and second-order paths for substitution, previously defined by the rate equation

$$-d[Pt^{II}ABLX]/dt = k_1[Pt^{II}ABLX] + k_2[Pt^{II}ABLX][Y],$$

are believed to be essentially S_N2 processes. Thus, small concentrations of a five-coordinate intermediate are first formed by one of the weakly-bonded groups becoming firmly attached, and the other weakly-bonded group being displaced. Such an intermediate can have either the configuration of a square pyramid or, if there is a rearrangement, that of a trigonal bypyramid. Of these, the latter seems the most consistent with kinetic data so far obtained. For the k_1 path, the following changes are envisaged. The positions above and below the plane are both occupied by solvent molecules S and the intermediate is formed by one of them partly displacing a fully coordinated group,

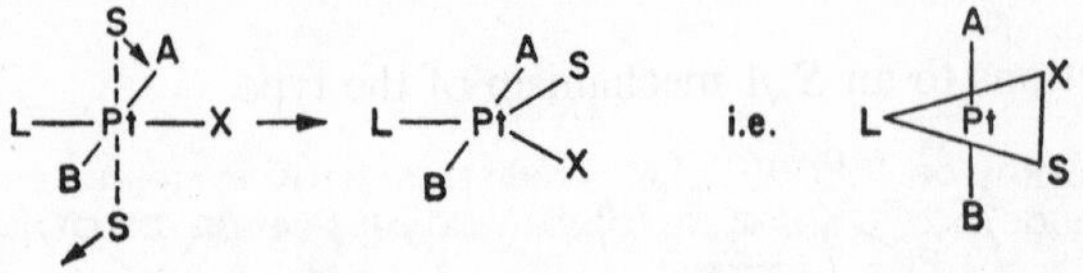

In this way a trigonal bipyramid is formed with L, X and S ligands in the same plane. The intermediate can then undergo a series of changes in which Y is substituted for S and S for X, following which, interactions with the solvent S result in the formation of the square-planar product,

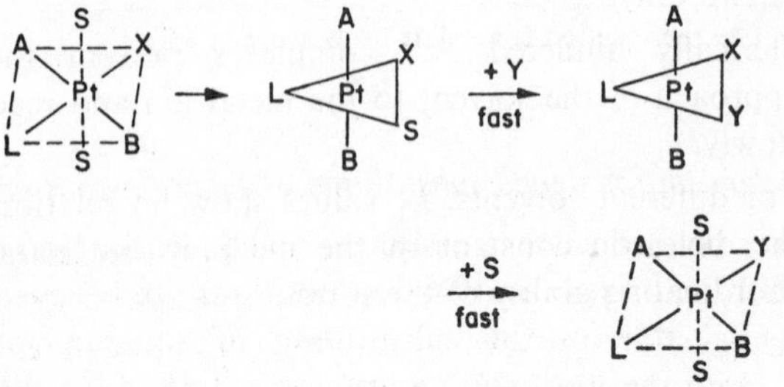

Similar changes are envisaged for the k_2 path where Y is, in the first instance, one of the weakly-bonded groups above or below the plane of the complex,

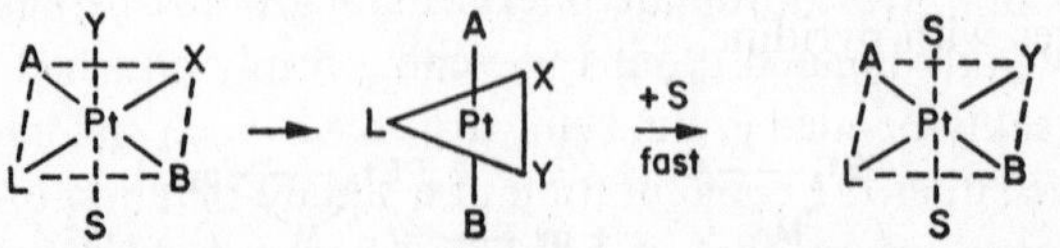

In both these mechanisms, the first step, in which a trigonal bipyramid is formed, is thought to be rate determining, and k_1 will therefore be independent of [Y] while k_2 will clearly show a

first-order dependence on the latter. The equivalence of X and Y in the transition complex is, in both cases, in keeping with relative rates obtained for systems in which X and Y are, in turn, varied.

Objections to an S_N1 mechanism of the type

A----X / Pt / L----B ⟶ A / Pt / L---B + X $\xrightarrow[\text{fast}]{+Y}$ A----Y / Pt / L----B

for path k_1 may be listed as follows:[5]

(a) For reactions in aqueous solution, k_1 is apparently little affected by the charge on the complex. Thus, the rates of hydrolysis of $PtCl_4^{2-}$ and $Pt(NH_3)_3 Cl^+$ differ by a factor of two only.

(b) Sterically hindered Pt^{II} complexes prevent the close approach of the solvent to the metal ion and react more slowly.

(c) For different solvents, k_1 values show no relationship to the dielectric constant of the medium, but parallel the coordinating ability of the solvent.

The effect of varying the central metal atom

In only one case has a direct comparison of rates of ligand substitution for analogous Ni^{II}, Pd^{II} and Pt^{II} complexes been possible (reference 4, page 256). This involves the replacement of the chloride ligand of *trans*-chloro-(*o*-tolyl)bis(triethylphosphine) complexes with pyridine,

PEt_3---Cl / M / *o*-tolyl---PEt_3 + py ⟶ PEt_3---py / M / *o*-tolyl---PEt_3 + Cl^-

5. See, e.g. H. B. Gray and R. J. Olcott, *Inorg. Chem.* **1**, 481 (1962).

In ethanol solutions at 25°C, rates are approximately 5,000,000 : 100,000 : 1 for M = Ni^{II}, Pd^{II} and Pt^{II}, respectively. The marked decrease in lability in going from Ni^{II} to Pt^{II} is in keeping with the S_N2 mechanism already outlined, since Ni^{II} is known to increase its coordination number more readily than Pt^{II}.

Substitution of the central metal atom

Of further interest are reactions in which the central metal atom of a complex is replaced by a second metal atom which is initially present as the free solvated ion. Reactions of the general type

$$MA_4 + {}^*M \rightarrow,$$

in which A is a monodentate ligand tend to be complicated, since a whole series of intermediate MA_n complexes (n = 1–4) are formed, and substitution of complexes which have bidentate and quadridentate ligands are generally easier to study.

The series of complexes of Co^{II}, Ni^{II} and Cu^{II} which are formed with bidentate and quadridentate Schiff-base ligands are particularly suitable for this type of investigation. With the bis(salicylaldehyde-*o*-phenyldi-imine)cobalt(II) complex,[6] the

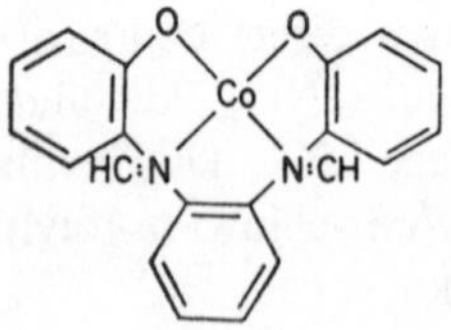

square-planar configuration, with one unpaired electron on the metal, can be confirmed by magnetic susceptibility measurements. In pyridine solutions the complex exchanges readily with free and

6. B. O. West, *J. Chem. Soc.* 395 (1954).

labelled Co^{II} (the acetate salt can be used) and the reaction is first order in both reactants. At 25°C, $k = 3{\cdot}3 \times 10^{-2}$ $\text{l mole}^{-1}\ \text{sec}^{-1}$, and from the temperature dependence, $\Delta H^{\ddagger} =$ 17 kcal mole^{-1}. That there is no reaction between the nickel(II) complex and free Ni^{II} over two days can be attributed to the stability of the square-planar d^8 configuration. In the corresponding reaction with the copper(II) complex and free Cu^{II}, $k = 4{\cdot}6 \times 10^{-3}$ $\text{l mole}^{-1}\ \text{sec}^{-1}$, and $\Delta H^{\ddagger} = 23$ kcal mole^{-1}. If, in these reactions, the central metal atom were completely freed by the breaking of four metal–ligand bonds, much higher activation energies would be expected. Instead, a mechanism in which there is stepwise dissociation or uncoiling of the ligand and simultaneous complexing to the exchanging metal ion seems more likely.

CHAPTER 14

SUBSTITUTION REACTIONS OF TETRAHEDRAL COMPLEXES

THE substitution of tetrahedral compounds of carbon can proceed by S_N1 and S_N2 mechanisms, although the latter are not of the limiting type in which five-coordinate intermediates are formed.[1] Investigations of the substitution of compounds of other Group IV elements, silicon, germanium and tin, have recently been made, and in these, S_N2 mechanisms are believed to be predominant. This is presumably because the latter elements can increase their coordination numbers more readily than carbon, an expansion of the outer octet of electrons being possible with *d*-orbital participation.

With tetrahedral complexes of transition-metal ions, an increase in coordination number to five is also possible, substitution reactions being generally very rapid and difficult to study. Exceptions are with those complexes in which the metal ion has a d^{10} electron configuration, when substitution is much slower and can be shown to be a limiting S_N1 mechanism.

The substitution of organosilicon compounds

With the synthesis of optically active organosilicon compounds of the general type R_3SiX, where R_3Si is, for example, the α-naphthylphenylmethyl group, it has been possible to demonstrate a high degree of stereospecificity in reactions in which X

1. C. K. INGOLD, *Structure and Mechanism in Organic Chemistry*, Cornell University Press, Ithaca, N.Y., 1953.

is replaced.[2] With essentially nucleophilic reagents, reactions may be divided into two groups depending on the nature of the leaving group. For good leaving groups such as Cl, F and OCOR (whose conjugate acids have pK_a values less than *ca.* 5), reaction proceeds with an inversion of configuration. Examples are the hydrolysis, methanolysis and lithium aluminium hydride reduction of R_3SiCl,

$$R_3SiCl \xrightarrow[\text{in ether}]{H_2O} R_3SiOH + HCl,$$

$$R_3SiCl \xrightarrow[\text{in pentane}]{CH_3OH} R_3SiOCH_3 + HCl,$$

$$R_3SiCl \xrightarrow[\text{in ether}]{LiAlH_4} R_3SiH + HCl.$$

For poor leaving groups such as H, OH and OCH_3 (whose conjugate acids have pK_a values larger than *ca.* 10), the predominant stereochemistry depends on the relative importance of a number of factors, and retention or inversion of configuration is possible in individual cases. Examples of reactions which proceed with retention of configuration are

$$R_3SiH \xrightarrow[\text{in } CCl_4]{Cl_2} R_3SiCl$$

and

$$R_3SiOCH_3 \xrightarrow[\text{in ether}]{LiAlH_4} R_3SiH.$$

These results support an S_N2 mechanism in which there is an expansion of the coordination number of silicon in the transition complex, rather than an S_N1 mechanism which would be expected to result in racemization. Evidence supporting an expanded octet (five-coordinate intermediate) mechanism has been obtained by studying the racemization of an optically active fluorosilane, R_3SiF, which has been shown to take place

2. L. H. Sommer, C. L. Frye, M. C. Musolf, G. A. Parker, P. G. Rodewald, K. W. Michaif, Y. Okaya and R. Pepinsky, *J. Am. Chem. Soc.* **83**, 2210 (1961).

without displacement of the fluoride ion.[3] Clearly this reaction does not proceed by an S_N1 mechanism since the fluoride is not even partly replaced by the methanol used to induce the racemization. At 25°C racemization of a 0·036 M solution of R_3SiF with 2·06 M methanol in t-butyl alcohol occurs at a rate $k = 0{\cdot}033$ sec^{-1}. The changes observed are consistent with the formation of a five-coordinate intermediate,

$$\underset{\text{I}}{\text{F}-\overset{\text{R}}{\underset{\text{R}''}{\text{Si}}}-\text{R}'} \rightleftharpoons \overset{\text{OMe}}{\text{F, R, R}', \text{R}''\ \text{Si}} \rightleftharpoons \underset{\text{II}}{\text{R}''-\overset{\text{F}}{\underset{\text{R}'}{\text{Si}}}-\text{R}},$$

where the silicon atom need not necessarily be coplanar with the F, R, R′ and R″ groups which are themselves planar. Structure (I) can be reformed by the movement of F and R′ groups towards the MeO^-, while the enantiomer in (II) results from the movement of R and R″towards the MeO^-.

The hydrolysis of triarylgermyl halides

Rate constants for the hydrolysis of a number of triarylgermyl halides,

$$R_3GeX + H_2O \rightarrow R_3GeH_2O^+ + Cl^-,$$

have been determined by conductance measurements.[4] The reaction medium was a solution of acetone–water or dioxan–water with ratios of solvent to water 5 : 1, 4 : 1 and 3 : 1 by volume. Acetone and dioxan were used because the reactants are sufficiently soluble in such solvents and because there is a large difference in dielectric constant. The results indicate a dependence on the amount of water present rather than the dielectric constant, thus, at 25°C and in 4 : 1 dioxan to water, the rate constant for the hydrolysis of $(C_6H_5)_3GeCl$ is 0·0090 sec^{-1}, while in 5 : 1

3. L. H. Sommer and P. G. Rodewald, *J. Am. Chem. Soc.* **85**, 3898 (1963).
4. O. H. Johnson and E. A. Schmall, *J. Am. Chem. Soc.* **80**, 2931 (1958).

dioxan to water, it is 0·0070 sec^{-1}. In 4 : 1 acetone, on the other hand, when the dielectric constant is 32 (compared to 11 in 4 : 1 dioxan) the rate constant is 0·0116 sec^{-1}. These results are contrary to an S_N1 mechanism in which there is formation of R_3Ge^+, and it has been concluded that reaction proceeds with the formation of a five-coordinate intermediate,

$$R_3Ge{-}X + H_2O \overset{\text{fast}}{\rightleftharpoons} R_3Ge\begin{smallmatrix} X \\ OH_2 \end{smallmatrix}$$

$$R_3Ge\begin{smallmatrix} X \\ OH_2 \end{smallmatrix} \rightarrow R_3GeOH_2^+ + X^-.$$

For a particular solvent, and with no extraneous salts added, the observed rate constants decrease in the order $R_3GeBr > R_3GeCl > R_3GeF$.

With the resolution of optically active germanium compounds of the type R_3GeX, where R_3Ge can be the α-naphthylmethylphenyl germyl group, it has been shown that reactions of germanium, like those of silicon, have a considerable stereospecificity.[5]

The solvolysis of organotin chlorides

The solvolysis reactions of organotin compounds R_3SnCl, where R is the ethyl, isopropyl, t-butyl or phenyl group, have been studied in ethanol, propan-2-ol and water dioxan solvents.[6] Solvolysis of tri-isopropyltin chloride with propan-2-ol is at a measurable rate, but in water-dioxan and in ethanol equilibrium is established too rapidly for conventional rate measurements, and this is true also of the tri-ethyl and tri-phenyl compounds in all solvents studied. When the alkyl group is the somewhat bigger t-butyl group, solvolysis rates are measurable in ethanol and in propan-2-ol, the resulting equilibrium lying further in

5. A. G. Brooke and G. J. D. Peddle, *J. Am. Chem. Soc.* **85**, 2338 (1963).
6. R. H. Prince, *J. Chem. Soc.* 1783 (1959).

favour of the undissociated halide. Because, (a) rates are very much dependent on the size of the alkyl and solvent groups (reactants having the largest groups react the slowest), and (b) solvolytic equilibria lie further in favour of the undissociated halide as the size and electron-releasing power of the alkyl group increases, S_N2 type mechanisms are believed to be applicable in these reactions. The converse of (a) and (b) would be expected if solvolysis proceeded by an S_N1 mechanism with intermediate formation of the " stannonium " ion R_3Sn^+.

The exchange of CO with Ni(CO)$_4$

The exchange of labelled carbon monoxide (at a pressure of 2·3 atm) with $Ni(CO)_4$ in n-hexane solutions is first order in complex and zero order in carbon monoxide.[7] It can be concluded that the reaction proceeds by an S_N1 dissociation mechanism,

$$Ni(CO)_4 \underset{k_{-1}}{\overset{k_1}{\rightleftharpoons}} Ni(CO)_3 + CO.$$

At 20°C, $k_1 = 5{\cdot}2 \times 10^{-3}$ sec^{-1}, and the activation energy is 24 kcal mole^{-1}.

That an S_N1 mechanism should be relevant, in this particular instance, is attributed to the d^{10} electron configuration of the Ni(0) atom. Because the *d*-orbitals are filled, there is a repulsive force on any attacking nucleophilic group, and an S_N2 mechanism is therefore unlikely. In addition, the geometry of the *d*-orbitals is such that, in a tetrahedral complex, only the $d_{x^2-y^2}$ and d_z^2 orbitals can π-bond readily to unoccupied *p*-orbitals on the ligands. There are, therefore, in effect, only two π-bonds to be distributed among the four ligand groups. More effective π-bonding is, on the other hand, possible for the three-coordinate planar intermediate formed in an S_N1 process, thus increasing the stability of such an intermediate.

It is of further interest (and related to the π-bonding) that a number of reagents, e.g. phosphines, arsines and pyridine will

7. F. Basolo and A. Wojcicki, *J. Am. Chem. Soc.* **83**, 520 (1961), and J. P. Day, F. Basolo and R. G. Pearson, *J. Am. Chem. Soc.* **90**, 6927 (1968).

react readily with $Ni(CO)_4$ with the replacement of two carbon monoxides. The rate of the first stage is the same as for CO exchange. Substitution of the remaining two carbon monoxides,

$$Ni(PR_3)_2(CO)_2 + PR'_3 \rightarrow Ni(PR_3)_2(PR'_3)CO + CO,$$

is possible but much slower. These reactions are also first order in complex and zero order in the incoming phosphine group.

The exchange of carbon monoxide with $Co_2(CO)_8$

In the exchange of labelled carbon monoxide (at a pressure of 1·4 atm) with $(CO)_3Co(CO)_2Co(CO)_3$ in toluene solutions, all eight carbon monoxide ligands exchange at the same rate.[7] Since the reaction is zero order in carbon monoxide, it must proceed by an S_N1 dissociation mechanism. It was originally suggested that the exchange required the formation of two equivalent $Co(CO)_4$ radicals. Such a mechanism can be discounted, however, since it would give rise to a square-root dependence on $Co_2(CO)_8$. Instead, a possible mechanism involves the initial dissociation of a bond to one of the carbon monoxide bridging groups,

$$(OC)_3Co(\mu\text{-}CO)_2Co(CO)_3 \underset{}{\overset{k}{\rightleftharpoons}} (OC)_3Co(\mu\text{-}CO)\text{------}Co(CO)_4 \underset{-CO}{\overset{\text{fast } +CO}{\rightleftharpoons}} (OC)_4Co(\mu\text{-}CO)\text{------}Co(CO)_4.$$

Since all eight carbon monoxides exchange at the same rate, it is necessary to assume free rotation about the cobalt atoms in this intermediate. At 0°C, $k = 1{\cdot}6 \times 10^{-3}\ \text{sec}^{-1}$ and the energy of activation is 16 kcal mole^{-1}.

The dichromate to chromate conversion

Both dichromate ($Cr_2O_7^{2-}$) and chromate (CrO_4^{2-}) are tetrahedrally coordinated chromium(VI) ions, the dichromate having

O—Cr(O)(O)—O—Cr(O)(O)—O (Cr—O—Cr angle 115°)

The latter form predominates in acid solutions, but is rapidly converted into the monomeric form when solutions are made alkaline, the overall reaction being

$$Cr_2O_7^{2-} + 2OH^- \rightarrow 2CrO_4^{2-} + H_2O.$$

The kinetics of this reaction have recently been studied by the stopped-flow method[8] using the spectral difference between reactants and products at 4750 Å. The rate equation is of the form

$$-d[Cr_2O_7^{2-}]/dt = k_1[Cr_2O_7^{2-}][OH^-],$$

and from the temperature dependence at $\mu = 0{\cdot}1$ M,

$$k_1 = 10^{6{\cdot}82} \exp(-5800/RT) \text{ l mole}^{-1} \text{ sec}^{-1}.$$

At 25°C k_1 is of the order 10^3 l mole^{-1} sec^{-1}. The mechanism is probably a bimolecular displacement of CrO_4^{2-} by OH^-,

$$\left[HO - - - Cr(O)_3 - - - O—Cr(O)_2—O \right]^{3-}$$

followed by the rapid neutralization of $HCrO_4^-$ by a second hydroxide ion.

At low alkalinity, e.g. pH 8, the hydrolysis reaction

$$Cr_2O_7^{2-} + H_2O \rightleftharpoons 2HCrO_4^-,$$

can be studied. Stopped flow and concentration-jump relaxation methods have been used,[8,9] the latter being applicable to the

8. C. D. Hubbard, P. Moore and R. G. Wilkins, *Proceedings 8th I.C.C.C.*, p. 286 (Edited by Gutmann), 1964.

9 J. H. Swinehart and W. Castellan, *Inorg. Chem.* 3, 278 (1964).

slower type of reaction system which is initially at equilibrium. In this method, sufficient concentrated chromate–dichromate solution is added to 3 ml of chromate–dichromate solution (containing an indicator) to change the total chromium concentration by 10%. Mixing of the two solutions is complete within 5 sec and subsequent colour changes of the indicator are recorded spectrophotometrically. The results have been explained by a mechanism

$$H^+ + In^- \overset{\text{fast}}{\rightleftharpoons} HIn \quad (1)$$

$$H^+ + CrO_4^{2-} \overset{\text{fast}}{\rightleftharpoons} HCrO_4^- \quad (2)$$

$$2HCrO_4^- \underset{k_{-3}}{\overset{k_3}{\rightleftharpoons}} Cr_2O_7^{2-} + H_2O \quad (3)$$

where In^- and HIn are the deprotonated and protonated forms of the indicator. Steps (1) and (2) are assumed to be fast, relative to step (3), which is reasonable, since protonation reactions are fast and generally in the region 10^{10} l mole^{-1} sec^{-1}. From the relaxation time, k_3 and k_{-3} are found to be 1·8 l mole^{-1} sec^{-1} and $2{\cdot}7 \times 10^{-2}$ sec^{-1}, respectively.

Oxygen-atom exchange with oxyanions

Of particular interest, here, are oxygen-atom exchange reactions in which oxygen atoms are exchanged between solvent water molecules and tetrahedral oxyanions of general formula XO_4^{2-}. A comparison with reactions of oxyanions having fewer oxygen atoms is also of interest.

In considering relative rates of exchange, the charge on the central metal atom would seem to be important since, for a series of ions $H_2SiO_4^{2-}$, HPO_4^{2-}, SO_4^{2-} and ClO_4^-, the rates decrease markedly in going from left to right. Thus, while exchange is rapid in the case of the silicate ion, it is slow for the phosphate ion, slower still for sulphate, and exceedingly slow with perchlorate. The latter is so slow that at room temperature the half-life for a 9 M perchloric acid solution is greater than 100 years.

With different oxidation states the lower ones generally substitute the faster. Thus exchange with sulphite ions is much faster than with sulphate ions, and, at the same acidity, nitrite ions exchange faster than nitrate ions. In the series of chlorine oxyanions, hypochlorite is known to exchange rapidly, while perchlorate exchanges only slowly, and rates probably decrease in the order $ClO^- > ClO_2^- > ClO_3^- > ClO_4^-$.

Detailed studies have been made in a number of cases. In the sulphate exchange, for example,[10] the rate law is of the form

$$\text{Rate} = k[SO_4^{2-}][H^+]^2.$$

Since SO_3 is a stable species an S_N1 mechanism of the type

$$2H^+ + SO_4^{2-} \overset{\text{fast}}{\rightleftharpoons} H_2SO_4$$

$$H_2SO_4 \rightarrow SO_3 + H_2O$$

$$SO_3 + H_2{}^{18}O \overset{\text{fast}}{\rightarrow} H_2S^{18}O_4,$$

has been suggested, exchange being accomplished by the fast second step. A similar second-order dependence on hydrogen-ion concentration is found with NO_3^-, ClO_3^-, BrO_3^-, CO_3^{2-} and NO_2^-, and seems to be characteristic of ions having a small highly charged central atom. With some of those, at least an S_N2 mechanism may be applicable, however. Thus it has been shown that chloride ions catalyse the oxygen exchange between water and NO_3^- and BrO_3^- ions.[11] The rate law for these reactions is

$$\text{Rate} = k[XO_3^-][H^+]^2[Cl^-],$$

which is consistent with a mechanism

$$2H^+ + XO_3^- \overset{\text{fast}}{\rightleftharpoons} H_2XO_3^+$$

$$H_2XO_3^+ + Cl^- \rightarrow ClXO_2 + H_2O$$

$$ClXO_2 + H_2{}^{18}O \overset{\text{fast}}{\rightarrow} H_2X^{18}O_3 + Cl^-.$$

10. T. C. Hoering and J. W. Kennedy, *J. Am. Chem. Soc.* **79**, 56 (1957).
11. M. Anbar and S. Guttmann, *J. Am. Chem. Soc.* **83**, 474 (1961).

If chloride ions function in this way, then by analogy, the uncatalysed reaction would be expected to proceed by a similar rate determining step

$$H_2XO_3^+ + H_2{}^{18}O \rightarrow H_2X^{18}O_3^+ + H_2O,$$

there being direct displacement of one water molecule by another.

When the central atom is larger or less highly charged, the rate law becomes first order in hydrogen-ion concentration, e.g.

$$\text{Rate} = k[XO_4^{n-}][H^+].$$

Such a rate law has been observed with ReO_4^-,[12] IO_3^-, OCl^- and OBr^-.

Finally, it should perhaps be mentioned that, for any one group in the periodic table, the rate of substitution seems to increase with the size of the central atom. This trend can be illustrated by reference to the halate ions. Thus, while chlorate ions exchange oxygen atoms with water at a measurable rate in acid solution at 100°C, the corresponding bromate reaction is measurable at 30°C. With iodate there is a rapid exchange at room temperature even in neutral solution.

12. R. K. Murmann, *J. Inorg. Nucl. Chem.* **18**, 224 (1961).

CHAPTER 15

ACID–BASE REACTIONS

IN recent years rate constants for a number of reactions in which there is transfer of a proton (but no accompanying change in electronic or molecular configuration) have been measured. Such reactions are almost invariably fast and relaxation techniques as used by Eigen and colleagues are generally required.[1]

Reactions in which one of the reactants is the hydrated proton, i.e. H_3O^+, are diffusion controlled with rates in the 10^{10}–10^{11} $l\ mole^{-1}\ sec^{-1}$ region (Table 37). The application of theoretical

TABLE 37

RECOMBINATION AND DISSOCIATION RATES FOR ACID EQUILIBRIA IN AQUEOUS SOLUTIONS AT ROOM TEMPERATURE

Acid–base reactions	k_1 (l mole^{-1} sec^{-1})	k_2 (sec^{-1})
$H^+ + OH^- \rightleftharpoons H_2O$	$1{\cdot}3 \times 10^{11}$	$2{\cdot}3 \times 10^{-5}$
$D^+ + OD^- \rightleftharpoons D_2O$	$8{\cdot}4 \times 10^{10}$	$2{\cdot}5 \times 10^{-6}$
$H^+ + F^- \rightleftharpoons HF$	1×10^{11}	7×10^{7}
$H^+ + SO_4{}^{2-} \rightleftharpoons HSO_4$	1×10^{11}	1×10^{9}
$H^+ + HCO_3{}^- \rightleftharpoons H_2CO_3$	$4{\cdot}7 \times 10^{10}$	$\sim 8 \times 10^{6}$
$H^+ + HS^- \rightleftharpoons H_2S$	$7{\cdot}5 \times 10^{10}$	9×10^{-3}
$NH_3 + H^+ \rightleftharpoons NH_4{}^+$	4×10^{10}	24

expressions for the diffusion of ions shows that the effective distance at which proton transfers occur is about 6 to 8 Å, i.e. there are intervening water molecules. According to Eigen, three stages may be considered for such reactions. (1) The formation of a

1. M. EIGEN and L. DE MAEYER. in *Techniques of Organic Chemistry* (Edited WEISSBERGER), 2nd edition, Vol. VII, part 2, p. 1031 (1963).

collision complex of acid and base. (2) The transfer of the proton through the hydration structure (by a Grotthus type mechanism) and its combination with the base. (3) The breaking of the hydration structure. The first stage (which is diffusion controlled) is the slowest and is rate determining, the second and third stages following very rapidly. The mechanism for such reactions may be illustrated in the following way

$$\begin{matrix} H \\ H \end{matrix}\!\!>\!O\!-\!H\cdots O\!<\!\begin{matrix} H \\ H\cdots B^- \end{matrix} \rightarrow \begin{matrix} H \\ H \end{matrix}\!\!>\!O\cdots H\!-\!O\!<\!\begin{matrix} H \\ (HB) \end{matrix}$$

Rate constants for the reverse dissociation reaction, k_2, in which a proton is set free are largely determined by the equilibrium constant. Thus the ratio of the rate constants k_2/k_1 for the reaction of H^+ with OH^- must be equal to the equilibrium constant which for water is $1{\cdot}8 \times 10^{-16}$ (the concentration of water is 55 M). With, for example, a 3+ transition metal ion, such as iron(III) which has a constant K of around 10^3 l mole^{-1}

$$Fe(H_2O)_5OH^{2+} + H^+ \rightleftharpoons Fe(H_2O)_6^{3+},$$

k_1 is of the order 10^{10}–10^{11} l mole^{-1} sec^{-1} so that k_2 for the dissociation reaction is of the order 10^7–10^8 sec^{-1}.

Reactions involving hydroxyl ions show similar features, (Table 38) where, again, proton transfer from the acid to the hydroxyl ion can occur by a Grotthus type mechanism

TABLE 38

RECOMBINATION AND DISSOCIATION RATES FOR BASE EQUILIBRIA IN AQUEOUS SOLUTION AT ROOM TEMPERATURE

Acid–base reaction	k_1 (l mole^{-1} sec^{-1})	k_2 (sec^{-1})
$H^+ + OH^- \rightleftharpoons H_2O$	$1{\cdot}3 \times 10^{11}$	$2{\cdot}3 \times 10^{-5}$
$NH_4^+ + OH^- \rightleftharpoons NH_3 + H_2O$	$3{\cdot}4 \times 10^{10}$	6×10^5
$HCO_3^- + OH^- \rightleftharpoons CO_3^{2-} + H_2O$	$\sim 6 \times 10^9$	—
$HCrO_4^- + OH^- \rightleftharpoons CrO_4^{2-} + H_2O$	$\sim 6 \times 10^9$	—

$$\begin{matrix} H \\ & \rangle O\text{---}H\text{—}O\text{—}A^{+} \\ & \; H \qquad\quad | \\ {}^{-}H\text{—}O & \qquad\quad H \end{matrix} \rightarrow \begin{matrix} H \\ & \rangle O\text{—}H\text{---}\left(\begin{matrix} O\text{—}A \\ | \\ H \end{matrix}\right) \\ & H \\ H\text{—}O \end{matrix}$$

Nuclear magnetic resonance techniques have been used to follow the proton exchange reaction between NH_3 and NH_4^+.[2] Two processes are predominant at pH 1·5–2·5, the first involving a direct exchange

$$H_3N\text{—}{}^{+}H\text{---}NH_3$$

$k = 11{\cdot}7 \times 10^8$ l mole^{-1} sec^{-1}, and the second proton transfer through an intervening water molcule,

$$H_3N\text{—}{}^{+}H\text{---}\underset{\displaystyle H}{\underset{|}{O}}\text{—}H\text{---}NH_3$$

$k = 1{\cdot}0 \times 10^8$ l mole^{-1} sec^{-1}. There is also a small contribution from reactions

$$NH_4^+ + H_2O \rightleftharpoons NH_3 + H_3O^+,$$

and at higher pH's, when the concentration of hydroxyl ions is appreciable, from reactions

$$NH_4^+ + OH^- \rightleftharpoons NH_3 + H_2O.$$

2. A. Loewenstein and S. Meiboom, *J. Chem. Phys.* **27**, 1067 (1957); S. Meiboom, A. Loewenstein and S. Alexander, *J. Chem. Phys.* **29**, 969 (1958).

APPENDIX I

CALCULATION OF EQUILIBRIUM CONSTANTS FROM FREE ENERGY VALUES

Equilibrium constants alone generally give no information regarding the rate of a chemical reaction. They do indicate the feasibility of a reaction, however, and can be used to predict the direction and extent to which a given reaction may proceed.

For gas-phase reactions, equilibrium constants can be calculated from standard free energy values. To evaluate free energies for compounds, the free energies of all elements in their standard states are assumed to be zero. The standard state of unit activity is that of a gas behaving ideally at unit (atmosphere) pressure, i.e. the gas at unit fugacity.

Example: To calculate the equilibrium constant for the $N_2O_4 \rightleftharpoons 2NO_2$ reaction at 25°C

The expression relating the free energy change for substances in their standard state to the equilibrium constant K is

$$\Delta G^0 = -RT \log_e K,$$

or at 298° K,

$$\Delta G^0{}_{298} = -1364 \log_{10} K$$

where ΔG^0 is in calories. From the tables below, the free energy change is obtained by subtracting the free energy of the reactants from the free energy of the products, i.e.

$$2 \times 12{,}270 - 23{,}400 = 1140 \text{ cal mole}^{-1}.$$

It follows, therefore, that

$$\log_{10} K = \frac{-1140}{1364}$$

and

$$K = 0{\cdot}12 \text{ atm.}$$

TABLE 39
STANDARD FREE ENERGIES AT 25°C (GAS PHASE)

Substance	ΔG^0 (kcal mole^{-1})	Substance	ΔG^0 (kcal mole^{-1})
H_2O	−54·6	NO	20·72
HF	−64·7	NO_2	12·27
HCl	−22·8	N_2O_4	23·4
HBr	−12·7	PH_3	4·36
HI	0·31	AsH_3	42·0
H_2S	−7·90	SbH_3	35·3
SO_2	−71·8	CO	−32·8
H_2Se	17·0	CO_2	−12·1
NH_3	−3·98	CH_4	−12·1

APPENDIX 2

CALCULATION OF EQUILIBRIUM CONSTANTS FROM STANDARD OXIDATION POTENTIALS

For solution reactions, equilibrium constants can be calculated using standard oxidation–reduction potentials. For each half-cell reaction, the equation

$$E = E_0 + \frac{RT}{nF} \log_e Q$$

is applicable, where E is the observed potential, Q the ratio of the activity of the oxidizing agent to that of the reducing agent, n is the number of electrons involved, and F the Faraday of electricity. The standard oxidation–reduction potential E_0 is for unit activities at 25°C, and is referred to the H_2/H^+ couple as zero. The sign of the potential (IUPAC convention) is the same as the charge on the electrode.

Example: To calculate the extent of the reaction of Cr^{2+} with Fe^{3+}

The thermodynamic equilibrium constant for the reaction

$$Cr^{2+} + Fe^{3+} \rightleftharpoons Cr^{3+} + Fe^{2+}$$

is expressed in terms of activities. Thus

$$K = \frac{a_{Cr^{3+}} a_{Fe^{2+}}}{a_{Cr^{2+}} a_{Fe^{3+}}}.$$

For the Cr^{2+}/Cr^{3+} half reaction

$$Cr^{3+} + e^- \rightleftharpoons Cr^{2+},$$

$E_0 = -0{\cdot}41$ V and $n = 1$, so that

$$E_1 = -0{\cdot}41 + 0{\cdot}059 \log_{10} \frac{a_{Cr^{3+}}}{a_{Cr^{2+}}}.$$

Similarly, for the Fe^{2+}/Fe^{3+} half reaction

$$Fe^{3+} + e^- \rightleftharpoons Fe^{2+},$$

$E_0 = +0{\cdot}771$ V and $n = 1$, so that

$$E_2 = 0{\cdot}771 + 0{\cdot}059 \log_{10} \frac{a_{Fe^{3+}}}{a_{Fe^{2+}}}.$$

Since, at equilibrium $E_1 = E_2$,

$$0{\cdot}41 + 0{\cdot}771 = 0{\cdot}059 \log_{10} \frac{a_{Cr^{3+}}}{a_{Cr^{2+}}} - \log_{10} \frac{a_{Fe^{3+}}}{a_{Fe^{2+}}},$$

hence

$$\log_{10} K = \frac{1{\cdot}181}{0{\cdot}059},$$

i.e.

$$K = 10^{20}.$$

When there is extensive complexing of anions with one of the metal ions, oxidation potentials often show a marked dependence on the nature of the acid medium. This is particularly noticeable for the $Ce^{4+} + e \rightleftharpoons Ce^{3+}$ reaction, where the $Ce(H_2O)_n^{4+}$ ion is extensively complexed, except in perchlorate solutions. Oxidation potentials are, for example, 1·7 V in 1 M $HClO_4$, 1·61 V in 1 M HNO_3, 1·44 V in 1 M H_2SO_4, and 1·28 V in 2 M HCl.

The variation in oxidation potentials as the inner coordination spheres of Fe^{II} and Fe^{III} are varied is illustrated in Table 41.

TABLE 40

STANDARD OXIDATION POTENTIALS IN ACID SOLUTION AT 25°C

Electrode reaction	Potential E_0 (V)
$Eu^{3+} + e \rightleftharpoons Eu^{2+}$	−0·43
$Cr^{3+} + e \rightleftharpoons Cr^{2+}$	−0·41
$V^{3+} + e \rightleftharpoons V^{2+}$	−0·255
$2H^+ + 2e \rightleftharpoons H_2$	0·00
$UO_2^{2+} + e \rightleftharpoons UO_2^{+}$	+0·05
$Ti(OH)_2^{2+} + 2H^+ + e \rightleftharpoons Ti^{3+} + 2H_2O$	+0·10
$Co(NH_3)_6^{3+} + e \rightleftharpoons Co(NH_3)_6^{2+}$	+0·10
$Np^{4+} + e \rightleftharpoons Np^{3+}$	+0·147
$Sn^{4+} + 2e \rightleftharpoons Sn^{2+}$	+0·150
$Cu^{2+} + e \rightleftharpoons Cu^+$	+0·153
$SO_4^{2-} + 4H^+ + 2e \rightleftharpoons H_2SO_3 + H_2O$	+0·17
$VO^{++} + 2H^+ + e \rightleftharpoons V^{3+} + H_2O$	+0·361
$I_2 + 2e \rightleftharpoons 2I^-$	+0·535
$MnO_4^- + e \rightleftharpoons MnO_4^{2-}$	+0·564
$Fe^{3+} + e \rightleftharpoons Fe^{2+}$	+0·771
$2Hg^{2+} + 2e \rightleftharpoons Hg_2^{2+}$	+0·92
$Pu^{4+} + e \rightleftharpoons Pu^{3+}$	+0·97
$VO_2^+ + 2H^+ + e \rightleftharpoons VO^{++} + H_2O$	+1·0
$Tl^{3+} + 2e \rightleftharpoons Tl^+$	+1·25
$Cr_2O_7^{2-} + 14H^+ + 6e \rightleftharpoons 2Cr^{3+} + 7H_2O$	+1·33
$Cl_2 + 2e \rightleftharpoons 2Cl^-$	+1·36
$HO_2^+ + H^+ + e \rightleftharpoons H_2O_2$	+1·5
$Mn^{3+} + e \rightleftharpoons Mn^{2+}$	+1·51
$Ce^{4+} + e \rightleftharpoons Ce^{3+}$	+1·61
$Co^{3+} + e \rightleftharpoons Co^{2+}$	+1·94
$Ag^{2+} + e \rightleftharpoons Ag^+$	+1·98
$S_2O_8^{2-} + 2e \rightleftharpoons 2SO_4^{2-}$	+2·01

[Reprinted, with permission, from W. LATIMER, *Oxidation States of the Elements and Their Potentials in Aqueous Solutions*, 2nd edn., 1952, Prentice-Hall, Inc., USA.]

TABLE 41

STANDARD OXIDATION POTENTIALS OF SOME IRON(II)–IRON(III) COUPLES

Electrode reactions	Potential E_0 (V)
$Fe^{3+} + e \rightleftharpoons Fe^{2+}$	+0·77
$Fe(PO_4)_2^{3-} + e \rightleftharpoons Fe(PO_4)^- + PO_4^{3-}$	+0·61
$Fe(CN)_6^{3-} + e \rightleftharpoons Fe(CN)_6^{4-}$	+0·36
$Fe(bipy)_3^{3+} + e \rightleftharpoons Fe(bipy)_3^{2+}$	+1·10
$Fe(phen)_3^{3+} + e \rightleftharpoons Fe(phen)_3^{2+}$	+1·14

APPENDIX 3

HYDROLYSIS OF METAL IONS AT 25°C

The extent of hydrolysis of some metal ions in perchlorate solutions is shown in Table 42. The K_h values given correspond to equations of the type

$$Fe^{3+} + H_2O \rightleftharpoons FeOH^{2+} + H^+,$$

and, unless otherwise stated, are at an ionic strength $\mu \to 0$. Hydrolysis constants vary appreciably with ionic strength; thus, for Fe^{3+}, $\log_{10} K_h$ is 2·22 with $\mu = 0$, 2·80 with $\mu = 0{\cdot}5$ M and 3·05 with $\mu = 3{\cdot}0$ M. [Further information, where available, can be obtained from *Stability Constants*, *Chem. Soc.* (*London*), Special Publication No. 17 (1964).]

TABLE 42

FIRST HYDROLYSIS CONSTANTS AND pH VALUES AT WHICH PRECIPITATION OCCURS WITH ~0·02 M CONCENTRATIONS OF METAL IONS

Ion	$-\log_{10} K_h$	pH for precipitation
Ce^{4+}	~0·7	2·7
U^{4+}	0·68	—
V^{3+}	2·9	—
Cr^{3+}	3·82	5·3
Fe^{3+}	2·22	2·0
Co^{3+}	1·75 ($\mu = 1{\cdot}0$)	—
Tl^{3+}	1·14 ($\mu = 3{\cdot}0$)	—
Pu^{3+}	6·95	—
Fe^{2+}	9·5	5·5
Co^{2+}	9·6	6·8
Cu^{2+}	7·5	5·3
Hg_2^{2+}	5·0	—
Hg^{2+}	2·49	—
Ag^+	11·7	7·5
Na^+	14·6	—
Mg^{2+}	11·4	10·5
VO^{2+}	4·77	4·3

APPENDIX 4

THE INTEGRATED FORM OF SIMPLE RATE EQUATIONS

In the rate equations below, a and b are used for initial reactant concentrations, and x is the change in reactant concentration(s) after a time t.

(1). First-order reaction,

$$\frac{-d(a-x)}{dt} = k(a-x), \text{ or } \frac{dx}{dt} = k(\text{a}-x),$$

$$\log_{10}(a-x) = kt/2{\cdot}303 + \log_{10} a.$$

(2). Second-order reaction, reactant concentrations equal,

$$\frac{-d(a-x)}{dt} = k(\text{a}-x)^2,$$

$$\frac{1}{(a-x)} = kt + \frac{1}{a}.$$

(3). Second-order reaction, reactant concentrations not equal,

$$\frac{-d(a-x)}{dt} = k(a-x)(b-x),$$

$$\log_{10}\frac{(a-x)}{(b-x)} = \frac{(a-b)kt}{2{\cdot}303} + \log_{10}\frac{a}{b}.$$

(4). Equilibrium reaction of the general type

$$\text{A} + \text{B} \underset{k_{-1}}{\overset{k_1}{\rightleftharpoons}} \text{C} + \text{D},$$

in which the concentration of the reactants at a time t is $a-x$ and of the products x,

$$\frac{-d(a-x)}{dt} = k_1(a-x)^2 - k_{-1}x^2.$$

At equilibrium

$$k_1(a-x_e)^2 = k_{-1}x_e^2,$$

where x_e is the change in reactant concentrations at equilibrium, and on integration,

$$\log_{10} \frac{x(a-2x_e)+ax_e}{a(x_e-x)} = \frac{2k_1at}{2{\cdot}303x_e}(a-x_e).$$

For other systems requiring a more complex treatment the reader is referred to *The Foundation of Chemical Kinetics* by S. W. Benson (McGraw-Hill), and *Kinetics and Mechanisms* by A. A. Frost and R. G. Pearson (Wiley, N.Y.).

APPENDIX 5

DETERMINATION OF THE RATE OF ELECTRON-TRANSFER BETWEEN Fe^{II} AND Fe^{III} FROM THE RATE OF ISOTOPIC EXCHANGE

For any one reaction solution, electron transfer is at a constant rate,

$$R = k[Fe^{II}][Fe^{III}],$$

since both the reactant concentrations are constant. In order to determine k, the Fe^{III} is labelled with a suitable amount of radioactive ^{59}Fe, and the approach of the equilibrium,

$$Fe^{II} + {}^*Fe^{III} \rightleftharpoons Fe^{III} + {}^*Fe^{II},$$

is measured. At suitable time intervals, samples of the reaction solution are taken, the Fe^{III} separated by chemical means (p. 27), and its activity determined using a simple Geiger counter assembly. The rate of the isotopic exchange is

$$\frac{-d[{}^*Fe_t^{III}]}{dt} = k[Fe^{II}][{}^*Fe_t^{III}] - k[{}^*Fe_t^{II}][Fe^{III}],$$

where $[{}^*Fe_t^{III}]$ is the amount of labelled Fe^{III} at a time t, etc. But

$$[{}^*Fe_t^{III}] + [{}^*Fe_t^{II}] = [{}^*Fe_\infty^{III}] + [{}^*Fe_\infty^{II}],$$

(with appropriate corrections here if the amount of radioactive decay during an experiment is significant), and substituting for $[{}^*Fe_t^{II}]$,

$$\frac{-d[{}^*Fe_t^{III}]}{dt} = k[Fe^{II}][{}^*Fe_t^{III}] - k[Fe^{III}][{}^*Fe_\infty^{III}] - k[Fe^{III}][{}^*Fe_\infty^{II}]$$

$$+ k[Fe^{III}][{}^*Fe_t^{III}].$$

Furthermore, assuming isotopic fractionation to be negligible,

$$\frac{[Fe^{II}][*Fe^{III}_{\infty}]}{[*Fe^{II}_{\infty}][Fe^{III}]} = 1,$$

so that

$$\frac{-d[*Fe^{III}_{t}]}{dt} = k[Fe^{II}][*Fe^{III}_{t}] - k[Fe^{III}][*Fe^{III}_{\infty}] - k[Fe^{II}][*Fe^{III}_{\infty}]$$
$$+ k[Fe^{III}][*Fe^{III}_{t}].$$

The latter can be rearranged,

$$\frac{d[*Fe^{III}_{t}]}{[*Fe^{III}_{\infty}] - [*Fe^{III}_{t}]} = kdt([Fe^{II}] + [Fe^{III}]),$$

and integrated to give

$$-\log_e([*Fe^{III}_{\infty}] - [*Fe^{III}_{t}]) = kt([Fe^{II}] + [Fe^{III}]) + \text{constant}.$$

The integration constant can be obtained for $t = 0$. Thus

$$-\log_e([*Fe^{III}_{\infty}] - [*Fe^{III}_{0}]) = \text{constant},$$

and if F is the fraction of exchange at a time t,

$$F = \frac{[*Fe^{III}_{t}] - [*Fe^{III}_{0}]}{[*Fe^{III}_{\infty}] - [*Fe^{III}_{0}]},$$

then it can be shown that

$$\log_e(1 - F) = -kt([Fe^{II}] + [Fe^{III}]).$$

In the general case, therefore, with reactant concentrations a and b,

$$\log_{10}(1 - F) = -\frac{kt}{2{\cdot}303}(a + b).$$

The latter is known as the McKay equation. It can be used whatever the mechanism, thus it holds equally well for the exchange between Ag^{I} and Ag^{II}, even though the rate is proportional to $[Ag^{II}]^2$.

APPENDIX 6

THE VARIABLE VALENCY OF SOME TRANSITION AND NON-TRANSITION ELEMENTS, OXIDATION STATES, STABILITY AND COLOUR

Scandium (d^1s^2)

Sc^{III}	Colourless.

Titanium (d^2s^2)

Ti^{II}	Rapid interaction with water. No Ti^{2+} aqueous chemistry.
Ti^{III}	Stable, reacts slowly with H_2O, O_2 and ClO_4^-. Violet.
Ti^{IV}	Stable, no simple Ti^{4+} ion exists, $Ti(OH)_3^+$ and $Ti(OH)_2^{2+}$ (or TiO^{2+}) are predominant in dilute $HClO_4$.

Vanadium (d^3s^2)

V^{II}	Normal form V^{2+}. Reacts with O_2 rapidly, and ClO_4^- slowly. $V(CN)_6^{4-}$ is inert, no definite evidence in many other cases. Violet.
V^{III}	Normal form V^{3+}. Also reacts wtih O_2. Dark blue in perchlorate solutions.
V^{IV}	Stable, hydrolysed to VO^{2+}. Bright blue.
V^{V}	Stable, in acid is hydrolysed to yellow VO_2^+. At higher pH's more extensive hydrolysis, polymeric forms.

Chromium (d^5s^1)

Cr^{II}	Normal form Cr^{2+}. Rapid interaction with O_2. Sky blue.

Cr^{III} Normal form Cr^{3+}. Stable, forms non-labile t_{2g}^3 octahedral complexes. Aquo ion is dark blue. Inner-sphere $CrCl^{2+}$ ions are green.

Cr^{IV} Kinetic evidence only for transitory existence in aqueous solution.

Cr^{V} Cr^{IV} is probably octahedral, while Cr^{V} probably has a tetrahedral structure, e.g. CrO_4^{3-}.

Cr^{VI} Stable, hydrolysed to CrO_4^{2-} which is yellow. Dimeric form $Cr_2O_7^{2-}$ in acid solutions.

Manganese (d^5s^2)

Mn^{II} Normal form Mn^{2+}. Stable, very faint pink.

Mn^{III} Disproportionates $2Mn^{III} \rightleftharpoons Mn^{II} + Mn^{IV}$, but is stable in presence of a large ($\times 25$) excess of Mn^{II} and ~ 3 N acid.

Mn^{IV} MnO_2 readily precipitated.

Mn^{VI} Hydrolysed to MnO_4^{2-}. Stable, but disproportionates in acid. Green.

Mn^{VII} Stable, hydrolysed to MnO_4^-. Maroon.

Iron (d^6s^2)

Fe^{II} Normal form Fe^{2+}. Stable, reacts very slowly with oxygen.

Fe^{III} Stable, Fe^{3+} ion pale violet, while $FeOH^{2+}$, $FeCl^{2+}$, etc. are yellow due to charge-transfer effects.

Cobalt (d^7s^2)

Co^{II} Normal form Co^{2+}. Stable, pale pink.

Co^{III} Hexaquo ion oxidizes water over period of about an hour. Blue-green colour. Wide range of t_{2g}^6 amine complexes which are generally a red colour.

Nickel (d^8s^2)

Ni^{II} Normal form Ni^{2+}. Only aqueous ion of nickel. Green.

Copper ($d^{10}s^1$)

Cu^I	Aquo ion can exist in only minute amounts, since it disproportionates $2Cu^I \rightleftharpoons Cu^{II} + Cu$. Complexed forms are more stable. These are coloured due to charge-transfer effects.
Cu^{II}	Normal form Cu^{2+}. Stable, bright blue.

Silver ($d^{10}s^1$)

Ag^I	Normal form Ag^+. Stable, colourless.
Ag^{II}	AgO solid stable and convenient source of Ag^{2+}. Disproportionates to Ag^I and Ag^{III}, the Ag^{III} oxidizing water. Brown colour.
Ag^{III}	Transient existence only.

Gold ($d^{10}s^1$)

Au^I, Au^{III}	No simple gold cations in aqueous solution.

Platinum (d^9s^1)

Pt^{II}	No simple aquo ions have been observed. Forms non-labile square-planar complexes.
Pt^{IV}	Non-labile octahedral complexes.

Mercury ($d^{10}s^2$)

Hg^0	Small concentrations, 10^{-7} l^{-1} mole, in H_2O at 20°C.
Hg^I	Stable, dimeric Hg_2^{2+} ions. Colourless.
Hg^{II}	Normal form Hg^{2+}. Stable, colourless.

Thallium (s^2p^1)

Tl^I	Normal form Tl^+. Stable, colourless.
Tl^{II}	Kinetic evidence only.
Tl^{III}	Normal form Tl^{3+}. Stable, colourless.

Tin (s^2p^2)

Sn^{II}	Stable, colourless.
Sn^{III}	Kinetic evidence only.
Sn^{IV}	Stable, but largely covalent. Extensive hydrolysis to $Sn(OH)_6^{2-}$, etc.

Antimony (s^2p^3)

Sb^{III}	Stable, shows some cationic behaviour, largely SbO^+ form.
Sb^V	Stable, covalent.

Cerium ($f^1d^1s^2$)

Ce^{III}	Stable, is colourless in spite of f^1 configuration. Absorbs in near u.v.
Ce^{IV}	Stable, hydroxy and polymeric species $Ce(OH)_2^{2+}$, $Ce(OH)_3^+$, and $CeOCeOH^{5+}$ are formed, less than 20% Ce^{4+} in 0·1 N $HClO_4$. Hydrolysed forms are yellow due to charge-transfer.

Europium (f^7s^2)

Eu^{II}	Colourless. Reacts rapidly with oxygen. Slow reduction of water. High absorption bond at around 320 mμ ($\epsilon \sim 520$) due to $4f \rightarrow 5d$ or $4f \rightarrow 6s$ transition.
Eu^{III}	Colourless.

Uranium ($f^3d^1s^2$)

U^{III}	Slowly oxidized by H_2O, rapidly by air. Red-brown.
U^{IV}	Stable, slowly oxidized by air to UO_2^{2+}. Green.
U^V	Transient UO_2^+, disproportionates to U^{4+} and UO_2^{2+}.
U^{VI}	Very stable UO_2^{2+}. Difficult to reduce. Yellow.

Neptunium ($f^4d^1s^2$)

Np^{III}	Stable in water, but oxidized by air. Purple.

Np^{IV}	Stable, slowly oxidized by air to NpO_2^+. Yellow-green.
Np^{V}	Stable NpO_2^+, disproportionates only in strong acid. Green.
Np^{VI}	Stable NpO_2^{2+}. Easily reduced. Pink.

Plutonium ($f^5d^1s^2$)

Pu^{III}	Stable to water and air. Blue-violet.
Pu^{IV}	Stable in 6 M acid, but at lower acid disproportionates to Pu^{3+} and PuO_2^{2+}.
Pu^{V}	Always disproportionates.
Pu^{VI}	Stable PuO_2^{2+}, fairly easy to reduce. Yellow-pink.

BIBLIOGRAPHY

For further reading the following books are recommended:

A. A. Frost and R. G. Pearson, *Kinetics and Mechanism*, 2nd ed., Wiley, New York, 1961.

K. J. Laidler, *Reaction Kinetics*, vols. 1 and 2, Pergamon Press, London, 1963.

C. N. Hinshelwood, *Kinetics of chemical change*, Clarendon Press, Oxford, 1945.

A. F. Trotman-Dickenson, *Gas Kinetics*, Butterworth, London, 1955.

S. W. Benson, *The Foundations of Chemical Kinetics*, McGraw-Hill, New York, 1960.

F. Basolo and R. G. Pearson, *Mechanisms of Inorganic Reactions*, Wiley, New York, 1958.

L. E. Orgel, *An Introduction to Transition-metal Chemistry*, Methuen, London, 1960.

E. F. Caldin, *Fast reactions in solution*, Blackwell's Scientific Publications, 1964.

The reader is also referred to the following review articles:

C. B. Amphlett, Isotopic exchange reactions between different oxidation states in aqueous solution, *Quart. Rev.* (London), 8, 219 (1954).

H. Taube, Mechanisms of redox reactions of simple chemistry, *Advances in Inorganic Chemistry and Radiochemistry*, ch. 1, vol. 1 (edited by Emeléus and Sharpe), Academic Press, New York, 1959.

D. R. Stranks, The reaction rates of transition-metal complexes, *Modern Coordination Chemistry*, ch. 2 (edited by Lewis and Wilkins), Interscience, New York, 1960.

R. G. Wilkins and M. J. G. Williams, The isomerism of complex compounds, ibid. ch. 3.

F. Basolo and R. G. Pearson, Substitution reactions of metal complexes, *Advances in Inorganic Chemistry and Radiochemistry*, ch. 1, vol. 3 (edited by Emeléus and Sharpe), Academic Press, New York, 1961.

J. Halpern, Mechanisms of electron transfer and related processes in solution, *Quart. Rev.* (London), 15, 207 (1961).

N. Sutin, Electron exchange reactions, *Ann. Rev. Nucl. Sci.* 12, 285 (1962).

BIBLIOGRAPHY

For further reading the following books are recommended:

A. A. Frost and R. G. Pearson, *Kinetics and mechanisms*, 2nd ed., Wiley New York, 1961.

K. J. Laidler, *Reaction kinetics*, vols. 1 and 2, Pergamon Press, London, 1963.

C. N. Hinshelwood, *Kinetics of chemical change*, Clarendon Press, Oxford, 1941.

A. F. Trotman-Dickenson, *Gas kinetics*, Butterworth, London, 1955.

S. W. Benson, *The Foundations of chemical kinetics*, McGraw-Hill, New York, 1960.

F. Basolo and R. G. Pearson, *Mechanisms of inorganic reactions*, Wiley, New York, 1958.

L. E. Orgel, *An Introduction to transition-metal chemistry*, Methuen, London, 1960.

E. F. Caldin, *Fast reactions in solution*, Blackwell's Scientific Publications, 1964.

The reader is also referred to the following review articles:

C. B. Amphlett, Isotopic exchange reactions between different oxidation states in aqueous solution, *Quart. Rev.* (*London*), **8,** 219 (1954).

H. Taube, Mechanism of redox reactions of simple chemistry, *Advances in Inorganic Chemistry and Radiochemistry*, ch. 1, vol. 1 (Edited by Emeleus and Sharpe), Academic Press, New York, 1959.

D. R. Stranks, The reaction rates of transition-metal complexes, *Modern Coordination Chemistry*, ch. 2 (Edited by Lewis and Wilkins), Interscience, New York, 1960.

R. G. Wilkins and M. J. G. Williams, The isomerism of complex compounds, *ibid*, ch. 3.

F. Basolo and R. G. Pearson, Substitution reactions of metal complexes, *Advances in Inorganic Chemistry and Radiochemistry*, ch. 1, vol. 3 (Edited by Emeleus and Sharpe), Academic Press, New York, 1961.

J. Halpern, Mechanisms of electron transfer and related processes in solution, *Quart. Rev.* (*London*), **15,** 207 (1961).

N. Sutin, Electron exchange reactions, *Ann. Rev. Nucl. Sci.* **12,** 285 (1962).

R. G. WILKINS, Kinetics and mechanisms of replacement reactions of coordination compounds, *Quart. Rev. (London)*, **16**, 316 (1962).

F. BASOLO and R. G. PEARSON, The *trans*-effect in metal complexes, *Progress in Inorganic Chemistry*, vol. 4, p. 381 (Edited by COTTON), Interscience, New York, 1962.

N. URI, Inorganic radicals in solution, *Chem. Rev.* **50**, 375 (1952).

J. O. EDWARDS, Rate laws and mechanisms of oxy-anion reactions with bases, *Chem. Rev.* **50**, 455 (1952).

and to the following:

The study of fast reactions, *Discussions of the Faraday Society*, **17** (1954), and Oxidation–reduction reactions in ionizing solvents, *Discussions of the Faraday Society*, **29** (1960).

AUTHOR INDEX

SUBJECT INDEX